Builder's Guide to Cosmetic Remodeling

Other volumes in the Builder's Guide Series
Builder's Guide to New Materials & Techniques *by Paul Bianchina*
Builder's Guide to Floors *by Peter Fleming*
Builder's Guide to Decks *by Leon A. Frechette*
Builder's Guide to Modular Construction *by Jonathan F. Hutchings*
Builder's Guide to Foundations & Floor Framing *by Dan Ramsey*
Builder's Guide to Barriers: Doors, Windows & Trim *by Dan Ramsey*
Builder's Guide to Wells & Septic Systems *by R. Dodge Woodson*
Builder's Guide to Residential Plumbing *by R. Dodge Woodson*
Builder's Guide to Change-of-Use Properties *by R. Dodge Woodson*

Builder's Guide to Cosmetic Remodeling

Chase Powers

McGraw-Hill
New York San Francisco Washington, D.C. Auckland Bogotá
Caracas Lisbon London Madrid Mexico City Milan
Montreal New Delhi San Juan Singapore
Sydney Tokyo Toronto

Library of Congress Cataloging-in-Publication Data

Powers, Chase M.
    Builder's guide to cosmetic remodeling / Chase Powers.
       p. cm.
    ISBN 0-07-050712-0 (hc).—ISBN 0-07-050717-1 (pb)
    1. Dwellings—Remodeling. I. Title.
TH4816.P68 1997
690'.837—dc21   96-49973
                                                                                        CIP

McGraw-Hill
A Division of The McGraw·Hill Companies

Copyright ©1997 by The McGraw-Hill Companies, Inc. All rights reserved. Printed in the United States of America. Except as permitted under the United States Copyright Act of 1976, no part of this publication may be reproduced or distributed in any form or by any means, or stored in a data base or retrieval system, without the prior written permission of the publisher.

1 2 3 4 5 6 7 8 9 0   BKP/BKP   9 0 2 1 0 9 8 7

ISBN (hc) 0-07-050712-0
     (pb) 0-07-050717-1

The sponsoring editor for this book was Zoe G. Foundotos, the editing supervisor was Scott Amerman, and the production supervisor was Claire Stanley. It was set in Garamond by McGraw-Hill's Professional Book Group Composition Unit, Hightstown, N.J.

Printed and bound by Quebecor Book Press.

McGraw-Hill books are available at special quantity discounts to use as premiums and sales promotions, or for use in corporate training programs. For more information, please write to the Director of Special Sales, McGraw-Hill, 11 West 19th Street, New York, NY 10011. Or contact your local bookstore.

Information contained in this work has been obtained by The McGraw-Hill Companies, Inc. ("McGraw-Hill") from sources believed to be reliable. However, neither McGraw-Hill nor its authors guarantee the accuracy or completeness of any information published herein and neither McGraw-Hill nor its authors shall be responsible for any errors, omissions, or damages arising out of use of this information. This work is published with the understanding that McGraw-Hill and its authors are supplying information but are not attempting to render engineering or other professional services. If such services are required, the assistance of an appropriate professional should be sought.

This book is dedicated to special people in my life. Big A and Little A give me my motivation. To all of you, and you know who you are, thanks for your support in this and all of my projects.

Contents

Introduction *xv*

1 Fixing foundations *1*
Pier foundations *3*
Cinder block foundations *5*
Brick foundations *6*
Solid concrete *8*

2 Materials for home exteriors *11*
Siding *11*
Roofing *14*
Paint *16*

3 Roofing for a new look *17*
Color *17*
Old roofs *18*
Check before you change *20*
False dormers *21*
Stoop covers *22*
Skylights *23*

4 Walls and ceilings *25*
Take a test *26*
Bricks *26*
Weathered barn boards *27*
Plaster *28*
Tongue-and-groove *37*
Paneling *39*

Bathrooms   *40*
Cedar shakes   *41*

5 Paint, stain, and wallpaper   *43*
Painting   *45*
Wallpaper   *48*
Staining   *51*
Tile   *51*
Borders   *52*
Murals   *54*

6 Flooring options   *55*
Carpeting   *56*
Carpet pads   *58*
Vinyl flooring   *59*
Break the rules   *65*

7 Plumbing propositions   *75*
Kitchen plumbing   *76*
Bathrooms   *81*

8 Electrical options   *95*
Switch covers   *96*
Outlets   *96*
Ceiling Fans   *96*
Lights   *97*
Exterior lighting   *100*

9 HVAC modifications   *107*
Exposed ductwork   *108*
Exposed piping   *108*
Bad baseboard units   *111*
Radiators   *111*
Outside equipment   *112*
New HVAC systems   *113*

10 Woodwork   *117*
Fingerjoint trim   *119*
Trim styles   *120*

Chair rail *121*
Crown molding *122*
Wainscotting *122*
Trim kits *124*
Prehung doors *124*
Other wood options *124*

11 Windows, doors, and skylights *127*
Windows *127*
Skylights and roof windows *131*
Exterior doors *133*
Doors for deck or patio *135*
Technical considerations *140*
Expenses *140*

12 Cabinets and countertops *143*
Counters *146*
Bathrooms *150*
Construction features *151*
Balance *154*
Lowball prices *156*
Installation problems *157*
Damage *158*

13 Accessories *161*
The exterior *161*
Foyers *162*
Living rooms *162*
Dining rooms *163*
Kitchens *163*
Bathrooms *164*
Laundry rooms *166*
Studies *167*
Bedrooms *167*

14 Kitchens *171*
Floors *171*
Walls *174*

Ceilings *176*
Windows *177*
Doors *179*
Cabinets *180*
Counters *181*
Plumbing *183*
Electrical *184*
Appliances *184*

15 Bathrooms *187*
Floors *187*
Walls *191*
Ceilings *197*
Plumbing fixtures *197*
Electrical Fixtures *198*
Windows and doors *198*
Accessories *200*

16 Living rooms *205*
Three types *206*
Selecting a style *208*
Material selection *213*

17 Family rooms *217*
Basement family rooms *219*
Main-floor family rooms *222*
Combination rooms *223*

18 Children's bedrooms *227*
Flooring *230*
Ceilings *230*
Walls *231*
Built-in units *232*

19 Adult bedrooms *235*
Fireplaces *235*
Terrace doors *236*
Skylights *236*
Ceiling fans *236*

Built-in fish tank *237*
Windows *238*
Ceilings *238*
Floors *239*
Walls *240*
Doors *242*
Built-in units *242*
Lighting *243*

20 Basements *245*
Basement upgrades *245*
Starting from scratch *247*
Basement floors *248*
Ceilings *249*
Walls *249*
Windows *250*
Doors *251*

21 Dining rooms *253*
Chair rail *253*
Room size *254*
Floors *256*
Ceilings *258*
Walls *259*
Doors *260*
Windows *262*
Lighting *265*
Identifying your path *265*

22 Laundry rooms *267*
Types of laundry rooms *267*
Closet laundries *268*
Basement laundries *269*
Garage laundries *271*
In-house laundries *272*

23 Sun rooms *275*
A room of many purposes *278*
Floors *279*

Walls *280*
Ceilings *280*
Lighting *281*
Plumbing *281*

24 Home offices *283*
Potential office locations *283*
Access *284*
Garages *286*
Basement offices *288*
Attic offices *288*
Existing rooms *289*
The requirements *291*

25 Home exteriors *293*
Windows *295*
Doors *295*
Railings *296*
Roofing *297*
Siding *298*
Gutters *300*
Shutters *301*
Lighting *302*

26 Decks and porches *303*
Existing porches *303*
Adding a porch *304*
Existing decks *306*
New decks *307*

27 Landscaping *311*
Landscaping plans *312*
Trees *313*
Foundation shrubs *313*
Flowers and groundcovers *313*
Timbers and mulch *314*
Walkways *314*

Garden settings *315*
Sprinkler systems *315*

Index *317*

About the author *325*

Introduction

Cosmetic remodeling is nothing new, but it is a type of work of which many contractors don't take full advantage. When people think of remodeling, they often envision walls being moved or stripped down to bare studs. They see a lot of inconvenient dust and mess. Homeowners often think of remodeling as an expensive option for improving their living conditions. This can be true, but it doesn't have to be. Cosmetic remodeling can be kept simple and inexpensive, yet still produce fabulous results.

I've worked as a builder and remodeler for all of my adult life, during which time I've run into countless remodeling situations. In addition to remodeling challenges, I've faced recessions, high interest rates, and a glut of contractors competing for the same jobs. To survive, I've come up with some special techniques, which I will share with you in this book.

There might have been a time when contractors could take out a few ads and get all the work they wanted. If there ever was such a time, I wasn't born early enough to enjoy it. Though my path as a contractor often has seemed like an uphill battle, the struggle has forced me to find new ways to outwit my competition. One of these methods involves offering customers a lot for a little. This doesn't mean that I don't make good money from the jobs I do. It simply means that I land the jobs that other contractors lose out of ignorance or indifference.

Any professional contractor knows that a complete kitchen remodeling job is likely to cost thousands of dollars. In this economy, homeowners might be afraid to make such an investment—but they could be quite willing to invest fewer thousands of dollars and still get a "new" kitchen. Bathrooms and kitchens are the two most popular rooms in a home to remodel. If you can find a way to give these rooms a new look for a little cash, you're onto something. This book will help you to reach that goal.

I won't be talking only about kitchens and baths. Every room you're likely to find in an average home is covered here. Not only will this book discuss interior improvements, it will dig into exterior improvements as well. You can find profitable information on all types of cosmetic improvements, ranging from roofing to landscaping, in this one handbook for professional contractors. In fact, there are 27 fact-filled chapters designed to help you put yourself into a higher tax bracket. A brief summation of them follows:

Chapter 1: Fixing foundations

Foundations can be ugly. When this is the case, it's hard to make any part of a house look good. What would you do if you were sprucing up a house that had an exposed cinder block foundation? You could brick it in, but this would cost a small fortune. Suppose you were to parge and paint the foundation? Now there's a cost-effective alternative to brick. This chapter deals with ways of adding a little class to ugly foundations. It also explains how foundation problems can be fixed in many cases with limited amounts of money.

Chapter 2: Materials for home exteriors

This chapter highlights the multitude of materials available for improving the outside appearance of homes.

Chapter 3: Roofing for a new look

Old roofs can really date a house. Replacing an existing roof with dimensional shingles can work wonders for the home's curb appeal. Building a false roof line can change the whole appearance of a house. Adding a small roof over a stoop can make a world of difference in the way a home invites people to visit it. This chapter divulges many good roofing ideas.

Chapter 4: Walls and ceilings

There are so many inventive things that can be done with walls and ceilings to change the flavor of a home. Why don't more remodelers move away from typical drywall ceilings and walls? Well, many of them don't want to take a risk. Some are too lazy, and others are not creative enough to try something new. However, once you read this chapter, you will not be bound to drywall alone.

Chapter 5: Paint, stain, and wallpaper

Paint, stain, and wallpaper are all common finishes for the interiors of homes. This is no surprise, but you will be surprised by all the fascinating things you can accomplish with these basic materials. In addition to giving you decorating ideas, this chapter shows you how to remove old wallpaper more easily and how to install new wallpaper so that it can be removed quickly and simply. You will also find

Introduction

many tips that can improve the appearance of a home for just a few hundred dollars.

Chapter 6: Flooring options

How many times have you limited the flooring options of your customers to carpeting and sheet vinyl? This is what most builders and remodelers do. Why? Get creative. Install a brick floor in a country kitchen. Put quarry tile in an entrance foyer. Think about using pumpkin pine flooring to create a rustic family room. Let this chapter show you all the options available for your customers.

Chapter 7: Plumbing propositions

Plumbing can become a complicated matter for remodelers. This is especially true when customers want a fresh look on a tight budget. What can you do? Did you know it's possible to buy some pedestal lavatories for less than the price of a vanity and top? You can. Tearing out a bathtub and replacing it can run into some serious money. Have the old tub refinished instead. This chapter concentrates on quick and easy plumbing alterations.

Chapter 8: Electrical options

Have you ever noticed how much difference a new light fixture can make in a house? I remember one house where I replaced an old hall light with a new one. The new light cast shadow designs on the ceiling and walls that mesmerized me. Something as simple as changing a few light fixtures can put a new twist on a home. Adding ceiling fans can also make a big difference. Read this chapter to learn all about electrical options.

Chapter 9: HVAC modifications

Few people think of HVAC modifications when doing a cosmetic makeover on a house. For many houses, adaptations to the HVAC are not needed. However, installing new registers and grills can produce pretty results. Getting rid of bulky radiators can allow for more flexibility in furniture placement. Check out this chapter to get in touch with HVAC possibilities.

Chapter 10: Woodwork

The woodwork in a house says much about the home's character. Replacing clam-shell trim with colonial trim could make a house look like a million bucks. Adding chair rail or crown molding turns a dining area into a formal dining room. Wainscotting can really reach out and grab you. This chapter provides you with all sorts of tips on woodwork.

Chapter 11: Windows, doors, and skylights

Few components of a home have as much effect on residents as windows, doors, and skylights. With the right use, these items can transform a dull, depressing house into a vibrant home. Adding a window or a skylight doesn't have to cost an arm and a leg. Read this chapter to see how you can give your customers a big bang for their buck.

Chapter 12: Cabinets and countertops

Cabinets certainly aren't cheap. But, they make a major contribution to the appeal of a home. A house with dark kitchen cabinets can be dreadful. Years of use tends to scar cabinets. Doing a complete replacement on a set of kitchen cabinets will cost thousands of dollars. But, refacing and refinishing is an affordable alternative. And, in this chapter you will learn all about the myriad ways of giving old cabinets a much needed face-lift.

Chapter 13: Accessories

Adding the right accessories to a house can make it a home. This chapter is packed full of simple, yet effective, accessories you can pitch to your customers for affordable improvements to their homes. You will enjoy the many illustrations which will make it easy for you to use this chapter as a sales tool when talking with your customers.

Chapter 14: Kitchens

Up to this point, we have been talking generically. For the rest of the book, we go on a room-by-room search for creative ideas and affordable cosmetic makeovers. Little things can mean a lot in a kitchen. Changing the face plate of a dishwasher from black to almond is one way to brighten up a kitchen. This is a cheap way to make an attractive change. Cutting in an octagonal window is no big deal, but it sure will change the appearance of a dark, boring kitchen. Some stenciling along the top of kitchen walls can set a theme for the rest of the kitchen to follow. If you read this chapter and free your mind to be creative, you will find countless ways to work magic in a kitchen.

The remaining chapters are similar in principle to Chapter 14. Each chapter deals directly with the type of rooms listed in the chapter title. Builders and remodelers are provided with good, affordable alternatives on their next jobs for each of these rooms.

Chapter 15: Bathrooms
Chapter 16: Living rooms
Chapter 17: Family rooms
Chapter 18: Childrens' bedrooms

Introduction

Chapter 19: Adult bedrooms
Chapter 20: Basements
Chapter 21: Dining rooms
Chapter 22: Laundry rooms
Chapter 23: Sun rooms
Chapter 24: Home offices
Chapter 25: Home exteriors
Chapter 26: Decks and porches
Chapter 27: Landscaping

This book is not your typical guide to remodeling that recommends "tearing it apart and putting it back together again." Many books concentrate on major remodeling jobs that can require customers to secure second mortgages on their homes. This is an intimidating and sometimes dangerous thing for homeowners to do. With house prices holding steady or declining, homeowners don't have the big hope of major appreciation they had several years ago. This doesn't mean that they are satisfied with what they have, however. Contractors who can show homeowners how to achieve new looks with minimal money probably will become the most successful ones in the field in the near future. You too can do this once you are familiar with the tactics and techniques in this book.

Take a few moments to thumb through the following pages. Look at the many illustrations and examples. One thing I believe is important in a book like this is real-world case histories, and I have included many of them for your review. Use them as a stimulus for your own creative thinking. When you read this book, it will be clear to you what customers want and need in today's remodeling market. It's not what it used to be. Take control of your future and reap the rewards made possible by following the suggestions contained in this guide.

Builder's Guide to Cosmetic Remodeling

1

Fixing foundations

Fixing foundations is a job most builders and remodelers would prefer to leave to someone else. There are times, however, when a remodeler or builder is forced to deal with an ugly or damaged foundation. Sometimes the problem is simple to fix, but the solution can be expensive. Knowing what to look for and what to do when questions about foundations arise is a skill all builders and remodelers should aim to acquire.

Homeowners can live in a house for years and not realize they have a foundation problem. Some damaged foundations go undetected until a contractor comes along to give an estimate for work to be done. When a contractor spots a foundation problem and brings it to the attention of the homeowner, it's never a pleasant surprise.

Leaking and damaged foundations are not the only issues a contractor faces. Have you ever had a customer call you and ask for a price for beautifying a foundation? I have, and it can be an unsettling situation in which to find yourself. The foundations under some homes are beyond cosmetic improvement, and need to be replaced. But many existing foundations can be given a new look at a modest cost.

It is said that beauty is in the eye of the beholder. I suppose that is true. During my many years as a contractor, I've been called to look at a lot of foundations. Most of the problems have been limited to leaks or cracks, but some have required giving a foundation a face-lift. For these jobs, I've come up with some creative ways to change the appearance of the foundations without going to enormous expense. Let me set the stage for an example that will put the creative side of your brain to the test.

Assume that you have been asked to provide suggestions and job estimates by a homeowner who considers an existing cinder block foundation to be ugly. The house is a rustic, A-frame design. The house, nestled in a grove of trees, has cedar siding and a lot of glass. Very little foundation is exposed on the front wall, but the side walls and back wall show a lot of block. Whoever laid the block was sloppy with the mortar joints, and your customer doesn't like the porous features of the ugly gray blocks. How would you suggest correcting this situation?

Your first thought might be to add a brick veneer to the foundation. This could be done, and it definitely would improve the home's appearance. Cost is a factor, however, and adding brick would be very expensive. What other ideas can you come up with?

The example we are studying is based on a real job that I was called out to in Virginia. My customer wanted to hide the unsightly cinder block inexpensively, while retaining the rustic qualities of the home and setting. One of the first things the customer said was that brick would not be acceptable, because of both its cost and appearance.

I knew that money was an issue, but I wanted to make my customer aware of various options, so I mentioned a stone veneer. The look of stone would blend well with the natural setting, and it would be a dramatic improvement over the existing foundation. Price was the stumbling block with stone, so I moved on to other options.

My next two suggestions were to install pressure-treated lattice or to parge the cinder block. Many of the new homes I was building at the time had parged-block foundations, and they looked good. The customer wanted to see what a parged foundation would look like before making a commitment. I provided addresses of various houses I had built with this type of foundation and waited to hear back from the customer.

My next call to the A-frame was to meet with the couple who owned the property. I was asked to bring a section of lattice with me to the job site. When I arrived, the wife had me place the lattice against the foundation to get a visual image of what the house might look like with lattice. While they were studying the look, I suggested planting shrubs and possibly some climbing vegetation, such as English ivy, to blend the lattice into the surroundings. They gave this option a lot of thought.

The two favored different options. The husband preferred having the block parged and painted, with a swirl pattern. He had seen this on one of the reference houses I had provided. The man's wife liked the idea of latticework and ivy. They were at a deadlock.

In the end, they compromised. My company was hired to parge and paint the front foundation wall and to install lattice on the side and back walls. The reasoning behind the couple's decision was that parging and painting the front section, which faced the road, would provide a slightly more formal appearance. Since the front wall was short, there was limited opportunity for a climbing vine to look its best. Ivy planted on the sides of taller walls would have plenty of room to grow. When the job was done, it looked fine.

A year or so later, I was called back to the house to remodel a bathroom. By that time, the ivy had grown considerably, and the place looked great. By training the ivy on the lattice, it was possible to keep the vine from invading the mortar joints and causing them to crack and crumble. Everyone was happy, and the job cost a fraction of what it would have if brick or stone had been used as a veneer.

As you can see from the example above, there are some inexpensive, creative ways to change the look of a foundation. Some solutions are even simpler. For example, I once was asked to provide advice on a brick foundation that had, in the owner's opinion, gone bad. The red brick foundation was dotted with white spots. After scrubbing and scrubbing, the homeowner gave up and called for professional help. I knew, even before seeing the job, what the problem and solution were.

When I arrived to give the homeowner an estimate, I was surprised to see how much efflorescence was on the bricks. The homeowner was concerned that the foundation repair would be extensive and expensive. After a quick walk around the house, I assured the homeowner that the problem would not be hard to rectify. The solution was to scrub the bricks with a mild solution of muriatic acid mixed with water in a 1 to 10 ratio. After a crew scrubbed and rinsed the bricks, they looked as good as new. My customer was delighted with both the look and the low cost.

Pier foundations

Pier foundations frequently are built for seasonal homes, houses along waterfronts, and mountain cottages. They are not typically used in urban neighborhoods. There have been builders, however, who have stuck houses up on piers in the most unlikely places. Many homeowners tire of owning the odd house on the block, and decide they want the piers hidden. This is where you, the builder/remodeler, come into the picture.

How do you hide a pier foundation? There are several ways to do the job, but many of them are expensive. Fortunately, there are cosmetic applications that can be used at minimal expense. How many ways can you think of to camouflage piers? Following are some of the options.

Brick or block

Brick or block foundation walls can be built to enclose a pier foundation. A footing must be installed before the walls can be built. Excavating around the edges for a footing will not be easy, and you can be sure the whole process will be very expensive. Few homeowners have the money or patience to authorize construction of a new foundation. Most homeowners will want some type of cosmetic fix-all that can be done quickly and will not cost an arm and a leg.

Lattice

I've seen lattice used to hide pier foundations. This can help, but it seldom gives the job a first-class look. Putting lattice in front of a solid foundation wall is one thing, but nailing it up so that light can shine through it from all directions is quite another. I've used lattice to hide the piers for decks, but I've never installed it to cover a pier foundation for an entire house. This doesn't mean that you don't have an option with lattice.

If you and your customer like the idea of lattice but don't want people to see between the slats, you can install a backing material before you put up the lattice. This backing could be plywood or wall sheathing, or even rigid insulation boards, which wouldn't be a bad idea for a house sitting up on piers. The insulation enclosing the foundation would improve the comfort of the home and would help protect plumbing pipes.

Nailing some backing material around a house's pier foundation is not a big job. Simply install some 2 × 4's between the piers to provide a nailing surface, and place your backing against the lumber before nailing it into place. Then you can apply the lattice over the backing, creating a closed-in look that works well with some types of homes.

Not all houses can support the look of lattice on their foundations. While lattice looks fine on a rustic home, it would look out of place on a traditional colonial home or other type of fancy house. You have to use common sense when picking a covering for a pier foundation.

Facades

The use of facades, such as half-bricks attached to backing, can produce a good-looking foundation screen. This, of course, is more expensive than lattice, but much less expensive than normal bricks and blocks. Facade materials come in all shapes, sizes, and colors. A visit to your local building materials store should produce lots of ideas and brochures for your next foundation job.

Cinder block foundations

Cinder block foundations probably are the most common type of foundation that people want to spruce up. A lot of houses have been built with exposed block foundations. Most people would agree that this type of foundation usually is not pleasing to the eye. The reason so many houses still have exposed block foundations is probably money. Adding brick to block increases the cost of construction considerably. To avoid this, some contractors and homeowners accept the fact that exposed block is the best option under the circumstances. As time passes and houses change owners, people decide to enhance the appearance of their homes by doing something with the foundation. But they rarely know what they want to do.

Just paint it

Some customers who complain about a block foundation might think that you should just paint it. I wouldn't advise doing this. I've seen a number of painted block foundations, and that's exactly what they look like, painted block foundations. Paint adds color, but little more. As a contractor, I would prefer to walk away from a job than to do work that will not look presentable. If someone asked me to paint a foundation, I would politely suggest other alternatives. Should the person insist on paint only, I would advise the customer to call a painting contractor.

Parge and paint

I'm very familiar with foundations on which parge-and-paint procedures are used to give an acceptable appearance at a low cost. I don't have any problem with offering to do this type of work, in most cases. Some exclusive homes would look strange with a parged and painted foundation, but such homes rarely have exposed block foundations to begin with.

Over the years, I've built a lot of new houses with parged and painted foundations. At the same time, I have installed a lot of brick foundations. The choice of which type of foundation to use was based on location, house style, and the price of the home. Many of my customers wanted low-cost housing that didn't look cheap. In these cases, I used block foundations and parged and painted them. After building dozens of houses on these foundations, I can say from experience that the work looks good and is durable.

When you parge a foundation, there are several proven patterns from which to choose. Parging can be done with a smooth, stippled, or swirled finish. Masons can use the trowel to create all sorts of unique textures such as a series of half-moons. I favor a finish with some texture in it. Swirl finishes are popular with my customers, as are rough textures. Once painted, usually in an earth tone, the finished foundations look appropriate.

The cost of parging and painting a block foundation is next to nothing when compared to the cost of brick or stone veneer. A parged foundation can be used effectively on a variety of house styles. Contemporary homes, Cape Cods, ranches, and even colonials can be built on parged foundations. Pigment can be added to the parge mix so that the process is done quickly and provides a uniform appearance. In my opinion, parging and painting is the best option for hiding ugly block foundations.

Brick foundations

Brick foundations normally don't require major cosmetic work, and most people are content with that type of foundation. Occasional cleaning or repair work can be necessary, however. While this work often is simple, to the homeowner it can seem like a monstrous job. Knowing what to do and how to do it is the key to working with brick foundations. Let me give you a few tips that might come in handy.

Cracks

Cracks, common in brick and block foundations, are generally caused by a settling foundation. Some settling is to be expected, but there are foundations that sink. If that happens, you've got a huge problem. Different types of soil support various amounts of weight (Fig. 1-1). When you patch a foundation that is sinking, you can end up with a disgruntled customer.

Foundations that crack still can last for decades. If water gets into the cracks and freezes during winter, more damage can be done. Bricks can be forced off a wall by ice forming behind them. Repairing the gaps and cracks between bricks is not a big job. Most homeowners could do the work themselves, if they knew how. But what happens if you repair the wall and a couple of months later the problem comes back? Your customer isn't going to be too happy.

When asked to repair cracks, you have two choices. If you are required to make the repairs right away, you should ask your customer to excuse you from liability for any continued settling. If the customer won't agree to that, you should suggest waiting a month or two to see

Tons/ft² of Footing	Type of Soil
1	Soft Clay – Sandy Loam – Firm Clay/Sand – Loose Fine Sand
2	Hard Clay – Compact Fine Sand
3	Sand/Gravel – Loose Coarse Sand
4	Compact Coarse Sand – Loose Gravel
6	Gravel – Compact Sand/Gravel
8	Soft Rock
10	Very Compact Gravel/Sand
15	Hard Pan – Hard Shale – Sandstone
25	Medium Hard Rock
40	Sound Hard Rock
100	Bedrock – Granite – Gneiss

1-1 *Safe loads by soil type.*

if the crack worsens. Apply a section of duct tape horizontally over the cracked area. By monitoring the tape for several weeks, you can tell if the foundation is still moving or if the crack is a result of previous settling. If the tape becomes twisted or split during your test period, it is likely that the foundation is sinking. When the tape remains intact, you can proceed with repairs that you might expect to last a while.

To repair cracks, remove all old mortar. Don't take shortcuts here. Leaving existing mortar in place under your repair will likely result in the repair work cracking or falling out. After removing the old mortar, it's a good idea to vacuum the cavity before installing repair grout.

Once the gap between bricks is clean, apply a liberal amount of grout material to the opening. A vinyl-based grout often is recommended because of its ability to overcome expansion and contraction. A pointing trowel should be used to force the grout into place. A striking tool can be used to finish the job with a seam that looks natural.

Pointing up

Pointing up bricks can enhance the appearance of a brick foundation. Old mortar joints sometimes turn sandy and discolored. Fix this problem by removing all existing mortar, doing a few bricks at a time. Remove mortar to a depth of about three-quarters of an inch. Don't stop until the remaining mortar is tight. Then brush the joints with a wire brush and wet them down with water so that they won't draw moisture from your mortar mix.

When you mix replacement mortar, it should be of a consistency that will slide slowly off your towel. Pack the mortar into the exposed joints, again using a pointing trowel. After you have done about a dozen bricks, go back over your work with a scrub brush to compact the mortar. Then, with a striking tool, put the finishing touches on your repair work. It's a good idea to dampen the new mortar once a day for two or three days after making the repairs.

Bad bricks

Some foundations suffer from bad bricks. These are bricks that have become damaged for one reason or another. Replacing individual bricks is not difficult. Matching replacement bricks with existing materials, however, can be difficult, especially if the home you are working on is an older one.

Assuming that you have found a suitable replacement brick, the process for installing it is simple. Wearing eye protection, use a wide mason's chisel to cut out the old brick and mortar. Most bricks can be broken out without much effort. Clean the cavity and vacuum it. Mix your mortar and wet down the wall cavity to prevent remaining bricks from robbing your mortar of water.

Soak the replacement brick in water. Apply mortar to the wall cavity on the bottom and both sides. You now can slide the replacement brick into place. Fill joint areas with mortar and then brush the joint. Use your striking tool to finish off the joint evenly.

Cleaning

The most common way of cleaning bricks is by using muriatic acid mixed with water. A mild mixture is usually all that is needed. What is a mild mixture? Try using 1 part acid with 10 parts water. This should be a strong enough solution for most cleaning jobs. Protect your eyes and skin when working with this type of material, and rinse the bricks after they have been scrubbed.

Solid concrete

Solid concrete walls are being used more often for single-family homes. These walls can be parged and painted, in much the same manner as cinder block. Some concrete foundations are smooth enough to allow painting without parging, but most aren't. The cosmetic recommendations offered for block foundations can be used with concrete. Defects in a solid concrete wall, however, are a bit more difficult to fix.

Fixing foundations

Dormant cracks can be filled with an epoxy-cement compound, but you shouldn't just shove this filling into an existing crack. The crack should be opened to a minimum width of ¼ inch, with a similar depth. A concrete-cutting saw should be used for this operation to ensure a clean, even cut. Clean the interior of the crack and allow it to dry. Then install grouting compound. Use a trowel to smooth out the sealant.

Cracks that continue to move require a different type of filling compound. An elastic-type sealant should be installed in active cracks. The size of the crack also should be made larger than for dormant problems. Make the repair crack 1 inch wide and ¾ inch deep. As with any type of product, refer to and follow manufacturer's recommendations at all times.

What you do with a foundation can have a tremendous effect on the overall beauty of a home. Think of a house with an ugly foundation as a person in formal party attire who is wearing torn sneakers. Foundations must be functional, but they also can be attractive.

2

Materials for home exteriors

A number of materials are available for home exteriors. Take siding, for example. There are at least half a dozen types of siding that you could be called upon to work with. Some types of siding are better than others for specific applications. This also is true for paint. How many types of shutter are there? Which type works best? Can you answer these questions? If a customer asks your advice in selecting a replacement for an entry door, what are you prepared to explain?

This chapter is a brief rundown of the types of materials available for exterior home improvements. Our purpose is to gain an overall understanding of what is available, and when and how it might be used. As you get further into this book, you will find more specific details on various applications. For now, we are going to seek an overview of exterior materials. Let's start with siding.

Siding

Siding is something that nearly every home has on it. Some types of construction, such as log homes, don't have a siding as such, but typical homes do. You might be asked to clean, paint, or replace siding. Before you start doing this type of work, you should know something about the products (Figs. 2-1 and 2-2).

Aluminum siding

Aluminum siding once was very popular. Its main selling point was that it didn't have to be painted. I can remember when my parents had aluminum siding installed. Today, aluminum siding is not used on many new installations. It is said to remain popular in certain parts of the country but, overall, it's been overtaken in popularity by vinyl siding.

Material	Care	Life, years	Cost
Aluminum	None	30	Medium
Hardboard	Paint	30	Low
	Stain		
Horizontal Wood	Paint	50+	Medium to High
	Stain		
	None		
Plywood	Paint	20	Low
	Stain		
Shingles	Stain	50+	High
	None		
Stucco	None	50+	Low to Medium
Vertical Wood	Paint	50+	Medium
	Stain		
	None		
Vinyl	None	30	Low

2-1 *Siding comparison.*

The life of aluminum siding is rated statistically at around 30 years, its cost is considered to be moderate, and it is fire-resistant. If you're looking for a siding that is easy to install over existing siding, aluminum siding is an option.

Aluminum siding dents easily, and the finish can chip and peel. Another problem with aluminum siding is that it tends to rattle when the wind blows. I would not install aluminum siding in new construction. I believe that vinyl siding is a better option.

Vinyl siding

Vinyl siding is good in many ways. The color is integral to the material, so it won't chip or peel. It can fade, however. Denting is not a problem with vinyl siding, and the cost of this siding is less than that of aluminum siding.

Many people pitch vinyl siding as being maintenance-free. It's true that the siding never needs to be painted, but it can require power washing to remove dirt and mildew. The life span of vinyl siding is said to be about 30 years. This siding installs easily over other types of siding, which makes it desirable for cosmetic contractors. Intense heat, such as from a fire, will melt the material.

Hardboard siding

I've used a lot of hardboard siding over the years. This siding usually cannot be installed over existing siding, but it is an inexpensive product for siding replacement. Like aluminum and vinyl siding, hardboard siding should last about 30 years. Due to its makeup, hardboard siding can suffer from moisture problems. Installation of this siding requires patience and attention to detail. Otherwise, wavy lines can be expected in the finished product. Some contractors say this siding can be stained, but most stick to paint for a good finish.

Material	Advantages	Disadvantages
Aluminum	Ease of installation over existing sidings Fire resistant	Susceptibility to denting, rattling in wind
Hardboard	Low cost Fast installation	Sometimes susceptible to moisture
Horizontal Wood	Good looks, if of high quality	Slow installation Moisture/paint problems
Plywood	Low cost Fast installation	Short life Sometimes susceptible to moisture
Shingles	Good looks Long life Low maintenance	Slow installation
Stucco	Long life Good looks in SW Low maintenance	Susceptibility to moisture
Vertical Wood	Fast installation	Barn look, if not of highest quality Moisture/paint problems
Vinyl	Low cost Ease of installation over existing siding	Fading of bright colors No fire resistance

2-2 *Advantages and disadvantages of different types of siding.*

Horizontal wood siding

Horizontal wood siding is popular in the areas where I work. It can be painted or stained with good results. Life expectancy is 50 years or more, and the cost of acquisition ranges from moderate (pine) to high (cedar). Wood siding looks good and wears well when cared for properly. If the siding is not painted or stained promptly, discoloration can be a problem. Moisture can attack wood, but that usually isn't a problem.

Vertical wood siding

Vertical wood siding is not nearly as common as horizontal wood siding. This is a type of siding that can look either very good or very bad, depending on the quality of the material and workmanship. Some jobs look like they belong on barns instead of on houses. Being wood, this siding can be stained or painted, and it might last for 50 years. Cost of acquisition is moderate, as a rule. Moisture might be a problem in some situations, but it usually isn't.

Wood shingles

Wood shingles, which can be difficult to install, are noted for slow installation. Another drawback of wood shingles is the high cost. In terms of advantages, shingles look good, enjoy a long life (probably 50 years), and are normally a low-maintenance material.

When wood shingles are not cared for, they don't wear well. Cracking and splitting can be a problem. Moisture and mildew can cause any siding to look bad, but I've seen this type of problem frequently with wood shingles. Except in special circumstances, I would avoid shingles as a siding material.

Roofing

Roofing can be part of any cosmetic contractor's job. Roofs get old and brittle. Some homeowners have roofing replaced just to get a new color or a new look. Most roofs are covered with asphalt shingles, making replacement easy. A new roof can be installed over an existing roof when asphalt shingles are used. As popular as asphalt shingles are, fiberglass shingles are gaining ground quickly.

Some roofing materials, including slate and tile, are much more expensive and more difficult to work with. Wood shingles and shakes are not particularly difficult to work with, but they are expensive and they don't always wear well. I've seen shake roofs with vegetation growing out of them.

Asphalt shingles

Asphalt shingles probably cover more roofs in the United States than any other type of roofing. These shingles are available in a variety of colors. They can be purchased in dimensional designs that produce an effect similar to that of wood shingles. For general applications, asphalt shingles are hard to beat. The life span of shingles varies, but asphalt shingles are usually rated for 15 to 30 years, with 20- to 25-year expectancies being common.

Fiberglass shingles

Fiberglass shingles look a lot like asphalt shingles. Being fairly new to the building scene, fiberglass shingles have not yet been through the test of time that asphalt shingles have. Many contractors swear by fiberglass shingles, and some swear at them.

As with asphalt shingles, fiberglass shingles are available in a wide range of colors and styles. They resist rotting and fire, and are durable. Most of the information I've read pertaining to fiberglass shingles has been good. Many of the local contractors and suppliers I've talked to, however, have some reservations about the use of fiberglass shingles in cold climates.

I know of two jobs in which fiberglass shingles were used that failed to last two years. In talking to my local supplier, I found that many contractors had complained of similar problems. The shingle manufacturer was quick to make good on the defects, but the evidence makes me leery of using fiberglass shingles in certain circumstances.

Cedar shingles

Cedar shingles are used on several types of houses. This type of roofing is expensive. The life of a cedar shingle is usually rated at 20 years, while cedar shakes can be rated for up to 50 years. If left unsealed, cedar shingles might rot, warp, or split. Cedar roofing that is not treated to make it fire-resistant has a poor fire rating.

Other roofing

Other roofing materials that you might encounter include slate, tile, and tin. These types of roofing are still being installed on some new buildings, but they are not common in most residential markets. If you run into these types of roofing, your work is more likely to be in the form of repairs rather than replacements.

Paint

Paint is one of the most frequently used materials for exterior improvement. Painting is a cost-effective way to alter the appearance of a home. Four basic types of paint are used regularly, with latex the leader. Each type of paint has some pros and cons, so let's look at them individually.

Latex paint Latex paint is, by far, the most user-friendly paint available. It cleans up easily and is durable. A fast drying time is another advantage to latex paint. In a pinch, latex can be applied over a surface that is slightly damp, although I recommend avoiding this practice. Latex paint is naturally mildew-proof, but it might not be compatible with existing oil-based paints.

Acrylic paint Acrylic paint is a type of latex paint, and there are many similarities between the two. Acrylic paint can be used on almost any type of surface, including masonry. Application of acrylic paint is similar to that of latex paint.

Alkyd paint Alkyd paint is a solvent-thinned, synthetic-resin paint. It shares many of the characteristics of older oil-based paints. One noticeable difference between alkyd paint and oil paint is that alkyd paint dries faster. If you want to cover an existing oil-based paint, alkyd paint is a good choice. It has excellent hiding power and levels better than latex.

Oil-based paint Oil-based paint continues to be used by some old-school painters, but few contractors continue to use it. It is used much more often for exterior work than for interior jobs. The only notable advantage to oil-based paint is its legendary durability. There are, however, many disadvantages. Oil paints dry very slowly. They produce strong odors, and cleaning up after a job can be a serious mess. Due to the long drying time, painters have to be concerned about unexpected rain and bugs ruining their work. In any situation, I would use alkyd paint instead of oil-based paint.

Paint, siding, and roofing are the major elements of exterior improvements. This is not to say that they are the only aspects of home exteriors to consider when doing cosmetic conversions. Shutters, for example, can make a big difference in the way a house looks. Replacing the front door of a house can change a home's appearance dramatically. Windows are another consideration when looking at the exterior of a home. All of these materials can play a part in remodeling work.

Subsequent chapters will provide specific examples of when shutters, new windows, and other types of exterior changes can and should be made. Our next stop along the way to successful cosmetic conversions is roofing. Turn the page for more detail on this part of the job.

3

Roofing for a new look

Roofing to give a home a new look can involve replacing existing shingles or building false roofs. The roof on a home can have a lot to do with the overall appearance of the property. Houses with simple, boring roofs can be dressed up with dormers and porch roofs. Switching from asphalt shingles to cedar shakes can make a huge difference in a house. Even going from dull shingles to the newer, dimensional style can take years off the image of a home.

Most homeowners don't consider building a new roof over the old one, but if you have a customer who wants a whole new look, this is one way to achieve it. Cost might be a problem, but there is no doubt that adding more pitch or putting in some special roofing features can transform a common roof into the talk of the town. For anything other than a typical shingle replacement job, you might have to prod your customers with good ideas. Feel free to use suggestions in this chapter to get more out of your next roofing call.

Color

The color of a roof can have a bearing on the overall appearance of a house. It also can have something to do with how warm or cool a house is. White roofs are more reflective than black ones, and thus keep an attic cooler in summer. I've never had a property owner ask me to replace a dark roof with a light-colored one to make a house cooler, but I have had requests for color changes, even when there was nothing functionally wrong with the existing roof.

Most contractors don't sit around their offices and try to figure out how to get more work changing roof colors. This type of work is not done often enough as an elective job to make marketing for it worthwhile. Since it is rare to get a call from a homeowner who wants a new color just for looks, it's even more important that you be prepared for the call.

You probably get dozens of calls a year from customers who want to add decks, remodel bathrooms, or build garages. If this is the type of work you are accustomed to bidding, you probably do all right in getting work. But when an unexpected call, such as one for a change in roof color, comes along, you might be so surprised that you stumble through the initial phone conversation. This can send a signal to the potential customer that you are not the right contractor for the job. If that happens, a job is lost.

Once you come to terms with the fact that some people will ask you to tear off a perfectly good roof covering and replace it with the same type of material, but in a different color, you are ready to move on to the next step. Go ahead and prepare yourself mentally for the day that you get the call to make a cosmetic color change.

What will you tell a customer who wants a new shingle color? A lot of contractors might start by offering to show the customer color samples. This is a good beginning, but don't let it be the end. You have an excellent opportunity to pitch additional improvements to a customer who is willing to spend money just for appearance. For example, you might convince the customer to upgrade to dimensional shingles. Depending on the circumstances, you might get the customer interested in a cedar shake roof. Since you will be installing a new roof anyway, this could be a good time to add dormers, a porch roof, or some other physical roof alteration. It is possible to turn a simple shingle job into a lot more work.

Old roofs

Old roofs sometimes rot out under the roof covering. This is something that you must cover clearly in your quotes and contracts. If you are taking a job where the roof covering and the roof sheathing will have to be replaced, not allowing for the sheathing work in your pricing will lose you more money than you planned to make. Inspect the jobs you estimate closely. Some customers might agree to pay extra for sheathing work that is unexpected, but others might try to force you to do the extra work without additional pay.

Old roofs with asphalt shingles are common, and fairly easy to deal with. Tin roofs can still be found on old homes, and there are slate roofs still in use. In addition to these types of roof coverings, you might run into tile roofs. Each type of roof presents its own

Asphalt or fiberglass shingle	4-in-12 slope
Roll roofing with exposed nails	3-in-12 slope
Roll roofing with concealed nails, 3" head lap	2-in-12 slope
Double coverage, half lap	1-in-12 slope

Lower slope: treat as flat roof; use continuous membrane system, either built-up felt/asphalt with crushed stone, or metal system with sealed or soldered seams.

Wood shingles may be applied on slopes as low as 3 in 12 exposures are condensed.

3-1 *Lowest permissible slopes for various roofing materials.*

challenges for cosmetic makeovers. A contractor's creativity can be tested in some of these situations.

Let's say that you are called to a job where the house is old. The roof has a low pitch and a tin covering (Fig. 3-1). The customer plans to have new siding and windows installed, and would like to make the roof look better. What types of suggestions would you make?

The most direct approach would be to remove the tin roof covering and replace it with modern shingles. This may also be the least expensive way of getting a new look. But will the new look be dynamic enough? Remember this old house has a low pitch.

You might find as you start to peel away the old tin that the roof sheathing or structure has been damaged by years of leaking. You should, of course, get into the attic and inspect for that type of hidden damage before you commit to a price. Let's assume that you do find some water stains on ceilings and some discolored wood in the attic. A little probing with a screwdriver has indicated that at least some of the roof structure will have to be replaced. Now what?

Because the homeowner is serious about making major cosmetic improvements with both siding and windows, you might be able to sell the idea of having you build a new roof over the old one. There will be more expense in the labor and material to build a new roof, but money can be saved on demolition and repair of the salvageable part of the existing roof. Shingle cost won't be much different, and the overall spread between the two options could be small enough to justify the added expense.

By building a new roof over the old one, you can make the house look like a completely different home. Just increasing the pitch will make a big difference. On the surface, this might sound a bit extravagant, but when you run the numbers, you may find that the overall cost between the two options isn't so great. More importantly, the homeowner will get more of a new look than was initially considered.

Check before you change

You should check local requirements, including covenants and restrictions in property deeds, before you agree to change any aspect of a home exterior. I know this is something most contractors don't do, but they should. You can act on the assumption that you are being hired to do a specific job and that as long as you do the work properly, you will get paid—but don't count on it. A homeowner who hires you to make a major cosmetic improvement might not be able, or even willing, to pay you, if the work you are doing violates some local prohibition. This may sound far-fetched, but it is not as off-the-wall as you might think.

Did you know that in some planned communities, homeowners can't have a type of mailbox other than the type provided for them? Can you imagine living in a neighborhood where the siding of your home must be painted with certain colors, and you don't have any say in the matter? The same types of restrictions can apply to roof structures and coverings. For example, a subdivision might require that roofs on ranch-style homes have no more than a 6-in-12 pitch. It may be further stated that all homes shall have dimensional shingles in earth tones or dark colors. The list of potential prohibitions could be very long indeed.

As a contractor, you might not expect to be held liable for violating some association rule or some deed restriction that you are not likely to know about. While you probably would not be found liable in court, that doesn't mean that getting your money would be easy. In an all-out battle, you probably would win in court, but the delay in getting paid and the cost of going after your money could eat up most, if not all, of your anticipated profit.

Put yourself in the position of an uninformed homeowner. You call a contractor to install a new roof for you. There are many types of roofing materials from which to choose (Fig. 3-2). The work is done, and the bill for the roofing is thousands of dollars. Before you pay the bill, you are contacted by a representative of your homeowners' association and told that your new roof violates recorded covenants and restrictions. You are told that the roof color must be changed within 30 days or you will face legal action. Are you going to pay your contractor promptly? Maybe, but a lot of people wouldn't.

Roofing Type	Minimum Slope	Life, Years	Relative Cost	Weight, Pounds/100 ft^2
Asphalt Shingle	4	15–20	Low	200–300
Slate	5	100	High	750–4000
Wood Shake	3	50	High	300
Wood Shingle	3	25	Medium	150

3-2 *Comparison of roofing materials.*

The covenants and restrictions of communities can border on the ridiculous. These limitations are designed to maintain a harmonious neighborhood that will not depreciate in value by the actions of some homeowners. The concept is good, but the reality can be absurd. I understand the reason behind conformity requirements, but I sure wouldn't want to live in a place where a group of people could tell me what color my house could be or what style mailbox I could use. Regardless of my personal feelings, many people do live in controlled subdivisions, and covenants and restrictions are something you should ask about before you begin work on the exterior of a home.

False dormers

Have you ever been asked to build false dormers on a house? I have, and it struck me as being strange the first time I got such a call. The customer said that most of the other Cape-style houses in his neighborhood had gable dormers on the front of their roofs. His didn't. I explained that dormers are usually installed to provide light and ventilation when the attic section of a Cape Cod is finished into living space. Then I asked if he planned to convert his attic to living space. He said that he wasn't, but that he wanted his house to look more like the other ones around him. I shook my head, but agreed to meet with him.

After meeting with the homeowner, I was surprised to learn that he didn't even want the dormers cut into his existing roof. All he wanted was to have the dormers built on top of his existing roof. The whole purpose was to create a more aesthetic appearance, with no desire for increased light or ventilation. It seemed silly to me to spend so much money for something that wasn't functional, but I agreed to build the dormers. The house did look better once the dormers were installed, but I would have at least made them functional had it been my house.

Potential Life Spans for Various Types of Roofing Materials

Material	Expected Life Span
Asphalt Shingles	15 to 30 years
Fiberglass Shingles	20 to 30 years
Wood Shingles	20 years
Wood Shakes	50 years
Slate	Indefinite
Clay Tiles	Indefinite
Copper	In excess of 35 years
Aluminum	35 years
Built-Up Roofing	5 to 20 years

All estimated life spans depend on installation procedure, maintenance, and climatic conditions.

3-3 *Potential life spans for roofing materials.*

Some people don't care much about function. They are driven by curb appeal and neighborhood approval. This was the case in my job with the false dormers. I might add, I got another job that entailed building false dormers about six months after my first one, and I haven't gotten a similar request since. You might never receive a request for false dormers, but don't be surprised if you do. Even if your customers aren't concerned with function, you should be prepared to talk about the life span of their new roofing (Fig. 3-3). This is a subject that comes up frequently.

Stoop covers

The stoop cover over a front stoop counts as a roof, so I'd like to share a quick story with you on this type of cosmetic improvement. A customer called my office several years ago and wanted an estimate on some outside improvements, mainly shutters for all windows on the front of her home. Putting shutters on a few windows didn't thrill me, but I went on the estimate to see what else might be available.

A stately colonial home stood on a 1-acre lot in a prestigious subdivision. The house was big, and it did need shutters to dress it up. A small utility room had been built on the right side of the home, and its independent A-roof helped to break up the boxy look of the house. But my first thought when I pulled into the driveway was that this house needed a roof over the front stoop. Not only would the roof be functional by protecting people from weather as they waited to get in the front door, it would break up the bleak front of the home.

After talking to the couple who owned the home about shutters, I expressed my opinion about the stoop cover. Before I left the house, I had a contract for the shutter and the little roof to be built over the stoop. I turned a peanut job into something a little more profitable, and created a front elevation that was much more pleasing to the eye. My customers were elated. This is the type of thing that you can do if you are observant enough to see opportunities and aggressive enough to take advantage of them.

Skylights

You might not think of skylights or roof windows as a part of roofing, but maybe you should. If you have a customer who is seeking a more modern look for the roof, either of these options can provide it. Installing roof windows in upstairs living space can be more effective in terms of light than adding a dormer, and costs less. Putting a skylight in the roof over a kitchen provides a new look for the roof and a brighter kitchen.

I've worked in construction for most of my life, and I've sold home improvements for many of those years, but I've yet to see a roofer pitch customers on skylights. In fact, few remodelers make any effort to sell roofing customers on skylights. I love it, because there's more work for me!

Many contractors believe that skylights belong only in contemporary homes or houses with vaulted ceilings. I disagree. Many contractors assume that living rooms, family rooms, and perhaps kitchens are the only rooms where skylights are appropriate. Again, I disagree. Think about it. Natural light is desirable in nearly any room of a house. Bathrooms with skylights are more cheerful. If the skylight is operable, it provides an excellent venting system for the moisture that so often destroys a bathroom.

Kitchens are ideal locations for skylights because of the light and ventilation they provide. Bedrooms can become more romantic when the stars and moon can be seen through a skylight. Putting a skylight in a formal dining room might not be acceptable, but almost any other room can benefit from the installation of a skylight.

Vaulted ceilings provide the perfect setting for skylights, but any room that has open attic space above it is a candidate for a skylight. Building a light box is not a big deal, and the benefits often outweigh the cost. If you pitch this idea to homeowners effectively, you will probably get a lot more work.

Old skylights can date a house. How many plastic bubble-type skylights do you see installed in new houses? In my part of the country, not many. A house that has small bubbles dotting its roof tends to strike people as an older home. Replace these bubbles with modern, operating skylights for maximum curb appeal and profit. If you put your mind to it, there is a lot that can be done with a roof to improve the appearance of a home.

4

Walls and ceilings

Walls and ceilings account for much of what makes a house look good or bad, old or new, and happy or sad. Is it safe to say that the walls and ceilings of a home are as important to a house as clothes are to people? I think so. Statements can be made by the material selections for walls and ceilings. Keep in mind that we are not talking about paint or wallpaper at this time, but actual construction materials used to create walls and ceilings.

What type of material is used more often than any other when applying a wall to rough framing? Drywall is the obvious answer. How about ceilings? The answer is the same, drywall. Just about every house being built today has walls and ceilings made of drywall. Why? It's cost-effective, easy to install, takes paint well, and is durable under normal conditions. All in all, drywall is an excellent choice for walls and ceilings.

If drywall is such a good product, why am I about to take aim at it? It's not that I don't like or don't use drywall. Quite the contrary; most of my building and remodeling projects are done with drywall. But there are times when the mold should be broken. Seeing the same old thing time and time again gets boring. If you have customers who want their homes to be special, you may have to deviate from normal construction practices and use other materials (Fig. 4-1).

Type	Thicknesses	Sizes	Uses
Regular	$1/4"$, $3/8"$, $1/2"$	4×6 to 4×14	Interior walls and ceilings
Moisture Resistant	$1/2"$, $5/8"$	4×6 to 4×16	Base for tile in bath, etc.
Fire Resistance, Type X	$1/2"$, $5/8"$	4×6 to 4×16	Fiberglass and additives in core for fire hazards or high heat areas

4-1 *Types of drywall available.*

Take a test

I'd like you to take a little test. It's a simple quiz and you won't be graded on it. The purpose of this exam is to see how open your mind is and how much experience you have in the use of nontraditional materials for walls and ceilings. Pick up a pencil and paper and make a list of all the various types of materials you can remember seeing used as wall materials. Do the same for ceiling materials. Next, list all the types of materials you can think of that might have a viable use, at one time or another, in the making of walls and ceilings. Once you have completed your lists, return to this section and compare your answers with the information that follows.

Did you make your lists? Some readers did and some didn't, I suppose, but you will improve your mental awareness by writing the answers down on paper. This is your last chance to take the test before I reveal answers and information. If you didn't take me seriously about creating a list, consider doing so now. For those of you who have completed your questioning or refuse to participate, we will jump right into the answers.

Bricks

How many times have you seen bricks used for interior walls? Assuming that you have seen them used, what was your opinion of them? Many people like brick walls on interior rooms, but some don't. Bricks tend to overpower some rooms, and they can make a room too dark. But the right mixture of bricks with painted walls can produce fabulous results.

One of the houses in which I grew up had a nice family room. The two side walls were covered in either paneling or drywall; I'm not sure which. One of the end walls consisted of glass sliding doors. The other end wall was made of bricks. The bricks were red, brown, and black. To this day, the brick wall stands out in my mind as being special. I can't tell you for sure if the side walls were paneled or not; I think they were, but I can see that brick wall as clearly as if I were bouncing a rubber ball off it today.

The brick wall made a lot of sense in that particular room. I was a rowdy boy, and being able to bounce balls off a brick wall saved wear and tear on the other walls. Not only were the bricks impervious to harm from me, they looked good and added a nice atmosphere to the room. Having a wall of glass at the other end of the room eliminated the problem of a dark, drab, living space. The combination of brick and glass worked well in that house.

Using bricks to build interior walls is not cheap, and there are limitations on where such a decorating plan will work well. Family rooms are one place where brick walls are both functional and attractive. I would consider using brick walls in a country kitchen, but I'd never put them in a formal dining room, unless the intention was to recreate a room of a certain period. Game rooms, family rooms, some kitchens, and perhaps studies are the types of rooms in which I can see bricks being used effectively as wall material. I don't think I would want a brick ceiling, however.

Weathered barn boards

Weathered barn boards can be purchased from people who tear down old barns. This type of lumber has a unique personality. It's not unusual to find few pieces of aged barn boards that look alike. Each board is likely to have its own wormholes, knotholes, or perforations from nailing patterns. This is what makes the material so desirable in selected types of rooms and in rustic homes.

Barn boards are rarely uniform in size and shape. This is part of their appeal. Cracks and other defects are common, and add to the mystique of the material. Granted, old barn boards are not for every home, but some rooms can really shine with the installation of barn-board walls.

I once had a customer who wanted to build a log home. The individual didn't want log walls for interior partitions, but neither did he want modern drywall to separate his rooms. After thinking about options for awhile, I came up with the idea of using old barn boards. The customer loved the idea, and the finished product looked great.

Remodelers can use barn boards to make a fast change in existing rooms. For example, you could take a family room in which drywall was the existing wall material and cover it with barn boards for a fast and easy modification. Light switches and outlets would have to be moved out or boxed in, but that would be about the only other change necessary. Some creative trim work around windows and doors can compensate for the thicker finished wall.

Many homeowners decide to finish basements into living space. I've done a lot of basement conversions, and one of them stands out in my mind as being the best. My customers lived in a fancy townhouse. Their basement was unfinished, with a walk-out door to their rear yard. The couple maintained professional status in their jobs and in the primary living section of their home. Wanting a place to kick back and relax, they decided to convert their unfinished basement into a private playroom. They asked my advice on what they should do and how it should be done.

After talking to the couple for an hour or so, I began to get a feel for what they were after. The couple liked to search ghost towns with metal detectors when they weren't working. Many of the mementos that they had accumulated from their hobby were western in nature. One demand was for a billiard room and poker table. Another prerequisite was a wet bar. A half-bath was considered to be a necessity, and anything else was open for discussion.

I thought about the space available and about the people who would enjoy it. While it seemed strange to create a western saloon in the basement of a posh townhouse, it seemed appropriate for the customers with whom I was working. The thought of expressing my opinion worried me a little. I was afraid I might be misreading the couple and blow the deal. But I went for it and laid out my suggestions.

As I told my story, I could see the customers' expressions change. They were obviously interested. When I was done, I got a standing ovation, literally. It turned out that I was the fourth contractor consulted for this job, and I was the first one to hit the nail on the head. My rustic western motif was a go, and a contract was signed.

During the conversion of the basement, I used only aged materials for walls and ceilings. Even the folding saloon doors were built out of old boards. The end result was like a set out of a western movie. All the conveniences of modern life were incorporated into a very rustic atmosphere.

I got a phone call from the couple a few weeks after the work was finished. They went on and on, telling me how relaxing their new hideaway in the basement was. As it turned out, they showed the special section of the house to several of their friends, and I was hired to convert other basements to special themes. In locations with adequate population, you might consider specializing in basement theme parks.

Plaster

Very few contractors use plaster for walls or ceilings in modern construction. This is not to say that plaster is not still worked with, and worked around, on many jobs. There are a lot of existing homes that have plaster ceilings and walls. Plaster can be a real pain to work with. Whether you are tearing it out, patching it, or covering it up, it is a material that often frustrates the most experienced workers.

My first home had plaster walls and ceilings. Both needed some cosmetic help. I used paneling and stippled drywall compound to hide the imperfections in my walls and ceilings. This worked pretty well. Many contractors hide plaster ceilings with grid systems (Fig. 4-2) and

Walls and ceilings 29

drop-in ceiling tiles (Fig. 4-3). Walls often are furred out and covered with drywall. The number of contractors who leave plaster as a finished product in remodeling jobs is quite low, I believe.

While few people care much for plaster these days, there are customers who insist on maintaining the integrity of a period home. If you are going to attempt to work with plaster as a finished product, make sure the person you have doing the work possesses the proper skills. Plaster can be very difficult to work with. Getting just the right look is not something just anyone can do. If I were taking

4-2 *Ceiling tile grid system.* Armstrong World Industries

Chapter Four

4-3 *Installing a ceiling tile.* Armstrong World Industries

on a plaster job for a finished product, I would subcontract a specialist to do the work.

It is much more common to hide plaster than it is to preserve it. Some remodelers rip out the existing plaster, furr out the ceiling joist or studs, and come back with drywall as a wall. This works just fine. There are also remodelers who don't bother with the demolition of existing plaster. They simply furr out over it and apply a new covering. In the case of ceilings, drop-in tiles are a convenient and common solution to hiding plaster (Figs. 4-4 to 4-6).

Other types of ceilings tiles, such as those that are stapled into place, are also used to hide plaster. The options available for ceiling tiles are much broader than they once were. If you're out of touch with ceiling tiles, you owe it to yourself, and your customers, to explore some of the new choices available (Figs. 4-7 and 4-8).

During cosmetic conversions on plaster ceilings, I've used everything from drywall mud and a potato masher (to create a pulled pattern in the drywall compound) to tongue-and-grove boards to create new looks. The advantage to a grid system is that you can level it with relative ease. If a joint compound is applied and swirled or pulled

Walls and ceilings

4-4 *A quality ceiling tile.* Armstrong World Industries

(Figs. 4-9 to 4-12), you don't have to worry about minor dips and bulges in existing plaster. But if you are putting drywall or some other type of rigid material in direct contact with an old plaster ceiling, plan on furring out the surface first.

Many houses that sport plaster ceilings have tall ceilings. A 9-foot ceiling is common in houses with plaster ceilings, and 10-foot ceilings are not uncommon. This allows room for the construction of a false ceiling. There have been times when I've framed new ceiling structures just below an existing plaster ceiling. This is often easier than trying to furr out the humps and bumps in plaster or old ceiling joists (Fig. 4-13).

When you are starting with tall ceilings, you have an opportunity to use exposed beams in your cosmetic work (Fig. 4-14). Beams work well in some kitchens and in many family rooms. If you have

4-5 *Ceiling tile samples.* Armstrong World Industries

4-6 *A completed tile ceiling.* Armstrong World Industries

Walls and ceilings

Material	Use
Wood-Fiber	Can be applied over plaster or drywall with adhesive
Mineral-Fiber	Drop-in panels which work with a grid system
Fiberglass	Drop-in panels which work with a grid system

4-7 *Ceiling tile applications.*

Type of Material	Cost	Features
Mineral-Fiber	Very expensive	Noncombustible
Fiberglass	Midrange price	May be fire-resistant
Wood-Fiber	Inexpensive	May be fire-resistant

4-8 *Ceiling tile comparison.*

4-9 *Swirling a ceiling with a brush.* Georgia-Pacific

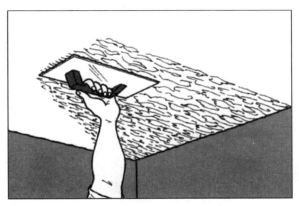

4-10 *Using a trowel to texture a ceiling.* Georgia-Pacific

4-11 *Stomping a ceiling with a stiff brush.* Georgia-Pacific

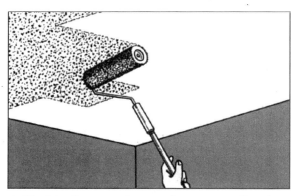

4-12 *Stippling a ceiling with a paint roller.* Georgia-Pacific

Cracked Ceilings
    Settlement in the building or foundation
    Vibrations in the building or foundation

Nail-Pops
    Nails pulling loose

Drywall Tape Coming Loose
    High humidity
    Improper installation

4-13 *Causes of ceiling defects.*

Walls and ceilings

4-14 *Exposed beams add to the charm of this kitchen.* Decora

exceptionally high ceilings with which to work, you might consider using beams as plant holders (Fig. 4-15).

Ceilings don't have to be flat, painted surfaces. Wood ceilings (Fig. 4.16) warm up a room. There are many good imitation wood products available for ceilings. Combining exposed beams with plank ceilings can have wonderful effects (Figs. 4-17 and 4-18).

Plaster walls can be difficult to work with. I've never seen one yet that wasn't uneven. I mentioned that my first house had plaster walls

4-15 *A plank-type ceiling that is accented by large beams.* Armstrong World Industries

and that I used paneling to hide the plaster. What I didn't tell you was what a mess I made of the first room I did.

When I started remodeling that first home, it was my first experience as a remodeler working with plaster. I bought wood paneling and installed it halfway up the living room wall. My idea was to have paneling halfway up, a chair rail, and then textured walls on up to the new ceiling. What I hadn't counted on was the wavy plaster walls. Being in a hurry and working in my own home, I didn't pay a lot of attention to the paneling as I installed it.

Walls and ceilings 37

4-16 *A kitchen with a plank ceiling.* Wood-Mode

When I had finished the first two walls, my wife came home. She gasped and put her hand over her mouth. Somehow, I didn't think she was taken aback with joy. I asked her what was wrong. She told me to come into the kitchen and look at the walls. When I did, it was obvious what had set her off. The paneling rolled across the wall with enough rises and falls to be a ski slope. I hadn't noticed the problem during installation. Being so close to the work surface, I was unable to see the defects. There was no way to ignore them when looking at them from a distance. Needless to say, I had to start over.

The solution to my problem was furring out the walls, which turned out to be a frustrating job. Fortunately, I was able to use the chair-rail trim to hide my little alterations, and the completed job was acceptable. This first experience with plaster taught me a valuable lesson. Avoid it whenever possible!

Tongue-and-groove

Tongue-and-groove (T&G) planking can be used on walls and ceilings. It's not cheap, but it does make a nice-looking job. The first new home I built had a lot of T&G planking in it. The house was an A-frame, so the T&G planking used to create the roof structure also

Chapter Four

4-17 *Random plank ceiling with beams.* Armstrong World Industries

served as my interior walls. Partition walls were made with drywall. The house had a sleeping loft, which had a T&G floor that doubled as a ceiling for the rooms under it. Many visitors commented on how much they liked the stained wood that were my walls and ceilings. This gave me the idea to incorporate T&G material into some of my remodeling jobs.

Since building my first house, I've installed T&G material for everything from ceilings to wainscot and access doors. The material is easy to work with, and it provides a lot of opportunity for creative

Walls and ceilings

4-18 *A beam ceiling in a kitchen.* Quaker Maid

images. Due to the thickness of T&G material, you could run into problems with some doors and windows, and you definitely will need to modify electrical outlets and switches, but the work is worthwhile.

You might be thinking that most of what we are talking about applies only to rustic types of homes. To a large extent, you're right. But the next chapter explores paint, wallpaper, stain, and murals, which can be used in other types of homes. I mention this now because I don't want you to feel that my views are one-sided.

Paneling

Paneling is not as popular as it once was. I can remember when any new house considered to have appeal had paneling. This is no longer true. Paneling still has its place, but the demand for it is down. Yet paneling is a quick way to change the appearance of a room, and it doesn't have to be expensive.

Suppose you have a customer who has just bought a used home and who wants to customize it, a common situation for remodelers. Let's say that the home has a variety of rooms, one of which has been targeted as a study for the man of the house. After talking to the gentleman, you understand that he has a love of the outdoors and enjoys fishing and all sorts of wildlife photography. He asks for your advice in dressing up the drab room which will become his personal sanctuary. What comes to mind?

Immediately, I think of paneling that is decorated with pheasants flying, grass blowing in the wind, and a forest in the background. Maybe a scenic of a waterfall and a lush mountain meadow would work better. In any event, you can offer your customer paneling that depicts any number of nature and wildlife scenes. This could be the perfect choice for changing the look of an old room. The work to install the paneling will go quickly, and the cost for labor and material will be affordable.

Now, what about the lady of the house who wants her painting studio to come alive with the look of a seashore? Once again, paneling could be the answer. Paneling is sold with a variety of scenes on it. Picking the right one will be up to your customers, but it is your job to let them know what is available.

There are some things that you have to look out for when working with paneling. First of all, stay away from the cheap stuff. Trying to make a job look good with flimsy paneling is an experience you don't need. Bid the job a little higher, with the understanding that the materials you are quoting are quality goods. The best tradesperson can only do so much with junky paneling, and it's never enough to make a good-looking job.

Dark paneling can be the downfall of a room. This is especially true in basement conversions, where paneling is so often used. Unless a room will receive a lot of natural light, stay away from dark shades of paneling. Of all the paneling problems I've seen over the years, having a room turn out too dark is the most common one.

Make sure you have a good base on which to apply the paneling. If you're remodeling, you probably will apply the paneling over existing wallboard. This should not present a problem. If you run into plaster walls, however, remember my mistake and check to see if the walls must be furred out first. I suggest that you glue and nail the paneling in place. This may seem like overkill, but I've found that the little extra work is worth the effort.

Bathrooms

When it comes to bathroom walls, there is one type of cosmetic application that I have found to be more trouble than it is worth. I'm sure you've seen it. The material I'm talking about is similar to paneling, except that it is made to look like ceramic tile or some other appropriate bathroom finish. My advice about this type of material is to avoid it. I've used it several times, at the direct request of customers, and I've never been happy with the way it turned out. After doing several jobs with it, I decided to decline any work where this type of material was specified. Maybe it's just me, but I can't see how this material could ever be installed to look like a first-class job.

Cedar shakes

You may have used cedar shakes on a roof, but have you ever installed them on an interior wall? I doubt if you have. Not many contractors have ever considered using roofing shingles as a wall covering. I came up with the idea of using cedar shakes as a wall covering after examining an old barn at a house I used to rent. The barn was sided with shingles, and I associated it with the weathered barn boards I had used in the past. Finding a customer who thought cedar shakes would look good on an interior wall took a while, but I did locate my first customer, and my crews installed the shakes in a country kitchen. The result was astounding.

The first house in which I ever installed shakes became a conversation piece. My customers enjoyed entertaining guests, so the job got a lot of exposure. Since most people like to show off their newly remodeled space, many visitors saw my creative touch in the new country kitchen I had created. The cedar shakes were a hit, and I got several follow-up jobs from my first installation.

Installing cedar shakes as an interior wall covering is not something that you are likely to do on a daily basis, but it could be just the creative edge you need to win a job or to make a name for yourself. The work is easy, and the cost is not outrageous. You can be almost certain that there will be little competition in the field, and your touch of rustic appeal can be all it takes to land several remodeling contracts.

We've talked a lot about walls in this chapter, and we'll talk about cosmetic wall coverings, such as paint, stain, and wallpaper, in the next chapter. To close us out here, I would like to encourage you to break away from tradition and to exercise your creative options when bidding and doing your next job.

5

Paint, stain, and wallpaper

Paint, stain, and wallpaper all are capable of producing some exciting new looks in older homes. Add tile, stenciling, and other decorating options to the list, and you can approach the cosmetic makeover of a house in much the same way that artists approach their easels. You become the home-improvement artist, with myriad choices at your fingertips.

If I had to guess, I'd say that paint and wallpaper are the items most often used to transform existing rooms into showplaces. This makes a lot of sense. The cost of a few gallons of paint isn't likely to break a homeowner's budget, and the result can be dramatic. Many homeowners tackle this type of job on their own, and some do very well with it. Others, however, fail rather miserably.

Oddly enough, I thought I wanted to become a professional painter when I was about 14 years old. To test my dedication, I offered to paint the balcony on my parent's home. Scraping the old paint off was more work than I thought it would be. Brushing new paint on the weathered wood took more coats than I had anticipated. Seeing insects trapped in the fresh paint frustrated me. By the time the job was done, I was sure I was not cut out to be a professional painter.

While many homeowners don't aspire to become professional painters, a lot of them think that they can do their own painting and achieve professional results. That is rarely the case. Painting is a lot more difficult than it looks. Sure, almost anyone can pick up a brush and roller and apply paint, but the result might leave a lot to be desired. As a contractor, you might get called in to finish projects that homeowners give up on. Even worse, you might be asked to right their wrongs.

Chapter Five

House painting isn't normally considered an art, but I have a tremendous respect for talented painters. After my first painting job on the balcony, I retired from painting until I built my first home. Being a mature builder, I was sure I could handle the paint work. Well, I did it, and it was acceptable, but if I'd been getting paid by the hour, my wages would have been too low to mention. Painting is just not something that comes naturally to me.

As bad as I am at painting, I'm worse with wallpaper. My first experience with wallpaper came when I decided to act as my own subcontractor on a job. The removal of wallpaper seemed so simple that I couldn't justify paying someone else to do it. The project was just a small half-bath, but it took me hours to strip the wallpaper. I got the job done, but I didn't save a dime, and I probably lost money, not to mention my patience. Never again would I be so foolish.

While I was ignorant enough to strip wallpaper in the bathroom, I was not stupid enough to think that I could hang new paper in the room. This was the only smart move I made with respect to the wallpaper on that job. Believe me, wallpaper is not as easy to work with as you might think.

Staining wood is a lot easier for me to do. I've done a lot of it, and remarkably, I've generally done it well. I've stained siding and trim with good results and only a few splinters in my hands. If my personal experience counts for anything, it proves that even average people normally can do a good job staining wood. There are, of course, some situations which require more skill than others. Still, staining is a lot easier than painting or wallpapering for me.

I've never tried to stencil a border around a room. It's probably just as well that I haven't, since I probably wouldn't be very good at it. But, stenciling is very popular, especially for kitchens, and is an area of cosmetic improvement to which contractors should pay attention.

Ceramic tile and I have had mixed results on jobs where I've acted in the capacity of installer. It's not cost-effective for me to do tile work myself, but I can do it. Normally, I rely on subcontractors for my tile work, and this works out well. Tile can be used effectively in kitchens and bathrooms, so it is a wall covering that deserves some inspection and respect.

Murals are similar to wallpaper in terms of what is required of an installer. I installed a mural in a bathroom once, but I'd never attempt to do it again. The bathroom was my own, and I was not thrilled with my performance. If I'd done the job for a customer, I'm sure I'd have torn out my work and hired a professional to clean up my mess. As you might gather, there are some aspects of construction for which I'm not the best man for the job. Murals, however, are popular in children's bedrooms, bathrooms, studies, and so forth. Any good paperhanger can satisfy your mural needs, and this is a market that you should take a serious look at.

Of all the various phases of work that go into building and remodeling homes, I'd say that painting is the one that most homeowners feel that they can do themselves. There have been many occasions when customers I was bidding work for wanted to save money and asked to do their own painting. I used to agree in order to get the jobs, but it didn't take long for me to realize that most people can't paint like a professional. When the paint work on a job is below industry standards, it reflects on the entire job. Keep this in mind if your next customer wants to paint the job to save some money.

When homeowners want to add a little sparkle to their homes, painting and wallpapering often come to mind. If this is the only work a property owner wants done, painting contractors probably will be called before general contractors. You might not get a lot of requests for painting jobs alone, but many of the jobs you bid will involve painting. Others will require staining or wallpapering. Your skill in advising customers in these areas could make the difference in whether or not you get the work.

Painting

Because painting is the most popular cosmetic conversion, let's talk about it first. There are a number of factors to consider when deciding whether to paint and what color or colors to use. Colors, both interior and exterior, play a vital role in the desirability of a house. Customers typically will have some color selections in mind when they call you, but it might be best for you to make some suggestions during your meeting with them.

Exterior colors are often pretty standard. A few people choose wild colors for the outside of their homes, but most stick with traditional colors. The style of a house can have a lot to do with what colors complement it. For example, a colonial home with gray siding, a red door, and brown shutters could look very good, while the same color combinations on a ranch-style home might not be attractive. Pink is rarely used an exterior color, but I know of one house that is painted a bright pink. The houses catches your attention, but not in a way that I find desirable.

A few miles away from the pink house is a brown house with a front porch. The house itself is fairly average-looking, but the posts that support the roof over the porch are real traffic-stoppers. Someone painted the posts with multiple colors, including yellow, pink, and white. I suppose this was done to make some statement, and I'm sure the pizza delivery people have no trouble identifying the house, but this type of paint work is crazy.

The exterior of a home should be painted with purpose. Colors should be chosen that work well with particular house styles. Covenants and restrictions should be checked to make sure what colors are acceptable. Another consideration might be to have a house blend in with a neighborhood or a natural setting. The final decision on color will come from your customers, but if you see they are going far out into left field, give them some advice.

Interior paint colors seem to spark something in some people. While most of the home interiors I've been involved with were fairly normal, I have gotten a number of unusual requests. When I build a house on speculation, I paint with neutral colors. Typical colors include off-white, cream, white, and some pastels. But I once had a customer request a purple bedroom. I remember a job where the customer wanted his study painted so that it would glow in black light. Painting a little girl's room pink is fine, but would you paint a formal dining room in the same shade? I don't think so.

The color of a room can affect how the room feels. Some color combinations make rooms seem small. Others brighten a room and appear to add square footage to the space. Green walls remind me of institutions, and gray walls remind me of schools. Stark white walls make me feel like I'm in a sterile zone, such as a hospital. Helping your customers choose good colors can be a big part of what makes a job successful. In subsequent chapters, we will discuss details for particular rooms.

While your customers should and probably will choose paint colors, it might be up to you and your painting contractor to recommend proper types of paint. If you are not familiar with common paint types, get familiar with them. Homeowners are taking more responsibility for learning about home improvements than they once did. It is likely that a customer will ask your opinion about making a paint choice. While you are not expected to be a professional painter, any good general contractor should possess basic product knowledge. Following is some information about paint that might come in handy during your next estimate.

Exterior paints

There are different types of paint that can be used on the exterior of a home. Latex, probably the most common paint used, cleans up easily and lasts a long time. It dries fast and resists mildew. Latex is normally a good choice, but there may be some occasions when latex paint is not compatible with a previous oil-base paint.

Acrylic paint is a type of latex. It dries very fast, and normally will cover any type of building material, including masonry work. It's as easy to apply acrylic paint as it is to use any other latex product.

Oil-base paint used to be the standard for exterior use. Latex has gained a lot of ground on oil-base products, and not nearly as many professional painters use oil-base paint as once did. The slow drying time for oil-base paint can be a problem if bugs are flying about or if rain should appear in the forecast. Other drawbacks include the difficulty involved with cleaning up oil-base paint and the strong odors that go with this type of paint. Durability is good with oil-base paint, but many professionals feel that latex products can hold their own in the test of time. Alkyd paint is the perfect choice for hiding blemishes. A synthetic-resin paint that is thinned with a solvent, alkyd paint shares characteristics with oil-base paint, but dries faster. Alkyd paint often is used to cover existing coats of oil-base paint.

Interior paints

When it's time to choose interior paints, there are many options available. You can get dripless paint for ceilings, acoustic paint for ceiling tiles (bet you didn't know this), and a host of other products for special applications. Since there are so many choices, let's look at them individually.

Latex is far out in front as the most popular interior paint. It goes on fast and cleans up easily. Latex can be bought with finishes ranging from flat to high gloss. The fast drying time of latex can make it possible for two coats to be applied in a single day. This type of paint is not as durable or as washable as alkyd or oil, but it does fine in most applications. If the surface to be painted is especially slick, latex is not a prime choice. Also, it is advisable to apply latex only to surfaces that have been primed with a latex or alkyd primer. (This is particularly true of plaster; without protection, the plaster's finish coat will absorb water from the latex paint, and thus be more prone to flake off.)

Oil-base paint has pretty much been eliminated from interior paint jobs. The odors associated with this paint are enough to put it out of contention. When oil-base paint is used, the odors must be considered both for their unpleasant smell and because they are flammable. Another disadvantage to oil is that it dries very slowly.

For customers who demand a stronger finish than latex can provide, you can offer alkyd paint. Unpainted surfaces should be primed with alkyd primer before the first coat of paint is applied. Alkyd paint will cover most any existing coating, and it is known for its ability to hide unwanted spots and blemishes. There are more fumes associated with alkyd paint than with latex, and alkyd dries more slowly. Solvents are needed to thin alkyd.

Urethane and polyurethane can be used on almost any porous surface or existing finish. It is common for trim to be finished with this type of coating to give it extreme durability. Applying these products is not always as easy as it might seem, but when you want something to resist grease, dirt, and abrasion, you can't beat them.

Epoxy paints are meant to be used on nonporous surfaces that have never been painted before, such as tile. This type of paint is very strong, and also expensive. There are few occasions when epoxy paint is used on common jobs, but it never hurts to know it exists.

Dripless paint is available for use on ceilings. This type of paint is expensive, but works fairly well. Acoustic paint also is available for ceilings where ceiling tiles are installed. Application of this paint is usually done with a sprayer, but special rollers also are available. The good thing about acoustic paint is that it is designed to cover ceiling tiles without affecting their sound-deadening qualities.

Texture paint can be used to hide problems with drywall finishes and other imperfections. Results of texture paint vary, but most produce a stucco-like look. Painters with experience and enough time can use texture paint to work wonders with rough walls and ceilings.

One-coat wonders, as I call them, are paints that are meant to require only one coat of coverage. The paint could be latex or alkyd. What makes the paint a one-coat paint? Extra pigment is added to the paint to increase its hiding power. This type of product works best when the surface being painted already has been sealed and has an existing color similar to the fresh paint being applied.

Wallpaper

Wallpaper is a very popular type of wall covering that normally is considered to be an upgrade over paint (Fig. 5-1). As with painting, many homeowners feel that they can hang their own wallpaper. Some can, but a lot can't do the work in a way that produces professional results. I've also known general contractors who thought they could do okay hanging wallpaper. Most of them couldn't, myself included. In my opinion, wallpaper should be applied by hands with lots of experience.

Wallpaper is available in a variety of forms, most of which are relatively easy to hang. I stress the word "relatively." Cosmetic contractors

Paint, stain, and wallpaper

5-1 *A wallpapered kitchen.* Armstrong World Industries

can use wallpaper to achieve stunning results. For example, a grasscloth wall covering can transform a pale, white room into a warm, inviting library or study. The use of a foil-type wallpaper is likely to be a major hit with teenagers who want to make their rooms rock and roll. Tough vinyl wallpaper can be installed in kitchens, bathrooms, playrooms, and other areas where wear and tear is likely. You can find wallpaper for all occasions.

One key to success with wallpaper is proper preparation. The wall surface should be clean and smooth. Walls that have never been painted should be prepared with an oil-base paint, then brushed over with sizing. This will make future removal of the wallpaper much easier. If you will be hanging paper over walls painted with latex paint, you should first treat the painted surface with a primer-sealer. Many professionals opt for this procedure in place of oil-base applications when preparing new walls for paper.

Some contractors feel it is acceptable to install new wallpaper over old paper. This can be done, but the existing paper must be tight and free of defects. Additionally, the old paper should be sealed and sized before any new installation takes place.

When wallpaper is applied directly to unprimed drywall, future removal can be a nightmare. Steam is usually used to loosen the adhesive on old wallpaper so that the material can be stripped off. When wallpaper is installed on a properly prepared surface, removal at a later date is much easier.

Wallpaper can be used in any room of a house. It might be run from a chair rail to a ceiling in a dining room. Or it might run from a chair rail to the floor (Fig. 5-2). Many upscale bathrooms are fitted with wallpaper (Fig. 5-3), and kitchens frequently are accented with wallpaper. Customers who want to add a little class to their walls often choose wallpaper over paint. Remember this on your next estimate.

5-2 *Wallpaper used on part of a wall with chair rail.* Congoleum Corporation

Paint, stain, and wallpaper

5-3 *Wallpaper used to set a theme in a bathroom.* Tubmaster

Staining

Staining woodwork in a house can make a big difference in the home's appearance. Replacing old doors with new doors that are stained can be like giving a house a new suit of clothes. Unfortunately, it is often necessary to replace existing trim with new trim in order to get the stained effect. This is a costly proposition. If you limit the stain work to public areas of a home, the cost is manageable and the results are outstanding.

Staining trim and doors is pretty easy; even I can do it successfully. While you probably won't get many jobs that entail ripping out new trim just to install stained trim, you might be able to work stained wood into your remodeling plans. For example, if you are giving a room a complete overall, replacing the trim won't be such a big deal. Upgrading to stained trim and doors is not always appropriate, but it's a good ace to have up your sleeve.

Tile

Tile can be used on walls to create lovely scenics or to produce a practical backsplash. For example, installing tile on kitchen walls in high-splash areas (such as behind a cooktop or sink) can reduce the chore of cleaning grease off walls, while producing a new look for the room (Fig. 5-4). Kitchens and bathrooms are the two most likely rooms to benefit from tile on the walls (Figs. 5-5 and 5-6). When you start considering cosmetic makeover for walls, you shouldn't rule out tile.

Borders

Borders have become popular, especially in kitchens. These borders, where walls meet ceilings, can be painted or created with wallpaper (Fig. 5-7). Adding a border to a kitchen can make a big difference in the room's overall appearance. It is not uncommon to find decorative borders in bedrooms, sun rooms, and even dining rooms. This is a small touch, but it can produce big results.

5-4 *Tile used as a backsplash in a kitchen.* Congoleum Corporation

Paint, stain, and wallpaper

5-5 *Extensive tile use in a kitchen.* HIP

5-6 *A designer bathroom tile.* Armstrong World Industries

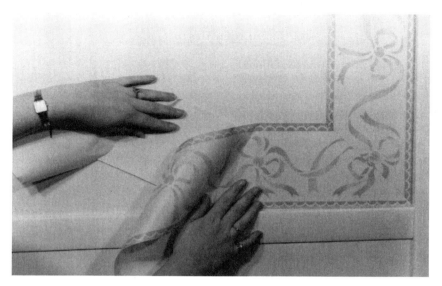

5-7 *Wallpaper stencil border.* Forbo, Lis King

Murals

Murals are generally installed on only one wall in a room. Options for scenes are diverse. You can provide your customers with a lush wilderness scene for the family room, a football scene for a bedroom, and a space scene for another bedroom. Murals allow people to customize their living space to suit their fantasies. Whether a customer's daughter is drifting off to sleep as fairies dance on her wall, or the customer relaxes in a new spa while gazing at a sunset over the ocean, murals definitely have their place in cosmetic makeovers.

We will talk more about specific wall coverings and ideas in subsequent chapters. Next, however, comes the lowdown on flooring options.

6
Flooring options

Flooring options for houses have never been more numerous. Why, then, do so many builders and remodelers limit their customers to choosing between carpet and vinyl flooring? Admittedly, carpet and vinyl are the two most common types of floors in use today, but they are far from the only choices available. Contractors who expand their list of options for customers often increase their business. Shouldn't you consider offering your customers something other than carpet and vinyl?

Have you ever considered putting a brick floor in a country kitchen? Maybe you should. Using old bricks to create a unique floor in a kitchen could earn your company a reputation as an innovator in creative remodeling. Bricks make a solid, durable floor that blends in well with country decor. We talked in Chap. 4 about using bricks on walls, so why not consider putting them on a floor?

The usefulness of bricks as a flooring material is limited, but how about tile? There are so many different types, sizes, and styles of tile from which to choose that you and your customers are sure to find something to suit any occasion. Tile floors are easy to clean, durable, and attractive. They can be slippery when wet, and dropped objects are more likely to break when hitting a hard surface like tile or brick, but tile certainly can be used in a number of places.

Foyers are excellent places for quarry tile. Wet boots and snow-covered clothing can be kept in the tiled area without fear that it will incur water damage. Tile with a baked-in texture reduces the risk of slipping. What could be a better choice? Slate or flagstone might be worth considering, but these materials become very slippery when wet, so I'd stick with textured tile.

Wood floors, both hard and soft, were in demand years ago, but you don't see much of them in modern construction. I suspect this is due to the high cost. While it is no longer practical to install hardwood flooring throughout a house, you can still dress up a special room with some attractive wood floors. (Pumpkin pine is a favorite of mine for rustic kitchen and dining areas.) The list of choices for wood flooring is a long one, and we'll talk more about it shortly.

Vinyl flooring is popular in bathrooms, kitchens, and laundry rooms. Many people use it in foyers, and some people use it in other rooms. The popularity of vinyl is well-earned. It is a dependable, cost-effective floor covering that is soft to walk on, easy to clean, and attractive when installed properly.

Carpeting rules the roost in home floor coverings. People have become spoiled by the warmth and comfort provided by wall-to-wall carpeting. When carpeting and vinyl first became popular, many people rushed to have them installed in their homes. In doing so, many fine wood floors were covered. It's not unusual to remove a piece of old vinyl or carpeting and discover a usable wood floor beneath it. With a little sanding and refinishing, a wood floor that was hidden for years can come to life with a new shine.

With so many types of flooring and such a vast opportunity for cosmetic improvements, how will you advise your customers? Are you going to point them toward the nearest carpet and vinyl dealers, or are you going to sit down with them and discuss more creative options? It's up to you, but I'd take the time to explain all of the products available to them. Doing so could result in more and better sales.

Carpeting

Carpeting covers more floors in more homes than any other type of floor covering. We all know what carpeting is, but not everyone understands the different types of carpeting available (Fig. 6-1). Do you know the difference between wool and nylon carpeting? I'm sure you recognize a difference in the material makeup, but do you know the advantages and disadvantages of these two types of floor coverings?

Wool carpeting tends to be very expensive (Fig. 6-2). This is partly because of its durability. Wool is resilient and abrasion-resistant. Another good feature of wool is that it's fairly easy to clean. The only downside, other than cost, is that it should be mothproofed.

Nylon, the strongest manmade fiber used in carpeting, gives wool a run for its money. Because it is extremely durable and resistant to abrasion, mildew, and moths, nylon is tough to beat. It even hides dirt well. If you have to point a finger at some undesirable feature, it is that nylon generates static electricity. This can be overcome by treating the fibers. It's hard to peg a price for nylon, because there is a significant range in cost.

Polyester:
    Bright colors
    Resists mildew and moisture
    Stays clean

Olefin:
    Very durable
    Resists mildew and moisture
    Very stain-resistant

Wool:
    Durable
    Abrasion-resistant
    Reasonably easy to clean
    Should be protected against moths
    Resists abrasion

Acrylic:
    Resists mildew
    Resists insects
    Remains clean
    Resists abrasion

Nylon:
    Extremely durable
    Resists abrasion
    Resists mildew
    Resists moths
    Remains clean
    Tends to create static electricity

6-1 *Carpet features.*

    Polyester is a favorite carpet material for people who like bright colors. Mildew and moisture don't normally present problems for polyester, but oil-based stains can be difficult to remove. For the most part, soiling is not a problem. Pricing for polyester carpeting falls in the midrange area.

    Acrylic carpeting resists abrasion, mildew, insects, and crushing. It's available in a wide assortment of colors, and it sheds dirt readily. The cost for this type of carpet is moderate. A drawback to acrylic is its tendency to pill or create fuzzballs.

    Olefin carpeting is the king of stain resistance. This carpet is extremely durable, nonabsorbent, and moisture-resistant. Cheap grades often crush in high-traffic areas, but more expensive versions stand up well.

Polyester	Moderate cost
Olefin	Prices vary
Wool	Expensive
Acrylic	Moderate cost
Nylon	Prices vary

6-2 *Cost comparison of carpet fibers.*

Loop-pile carpeting is a common type of carpet. Do you know why it's called loop-pile? Most modern carpeting is stitched to backing. If the fibers that are stitched into place are left uncut, the result is loop-pile carpet. When the fibers are cut, you have cut-pile carpeting. Unless your customers want to get really technical on the subject of carpeting, you should have enough product knowledge to engage in a meaningful conversation. If customers want more details, invite them to talk with your carpet supplier or installer.

Carpet pads

Carpet pads come in different styles, price ranges, and qualities. When carpeting is being installed, the padding beneath the carpet is a vital component of the floor-covering system. The very best carpet installed over a cheap pad will not have as long a life as it should. Most people, including many contractors, are not aware of the important role padding plays in terms of carpet performance.

When I was first getting started as a home builder, a carpet supplier stressed to me the role padding plays in carpet installation. He told me that a customer should invest in the best padding possible, even if it means that they have to settle for a less expensive carpet. I wasn't sure whether to believe this or not, so I asked for proof, and I got it.

The supplier pulled three different samples of carpet padding off his racks. Then he placed a sample of carpet over the first piece of padding and asked me to stand on it. I was wearing boots with heavy lug soles, so I made quite an impression on the carpet. After standing on the carpet for maybe a minute, I was instructed to step down and observe how long it would take for my footprints to disappear. Once they were gone, the supplier moved the same piece of carpet to the second pad and had me duplicate the test procedure. I did the same test on the third piece of padding.

The testing I did that day proved to me that the supplier wasn't lying. As the grade of the padding got better, my footprints disappeared more quickly. There was a noticeable difference between the cheapest pad and the most expensive. Keep in mind, the same carpet sample was used for all three tests. Since that day, I've been a firm believer that padding plays a major role in the life and look of carpeting.

Based on research, comments from suppliers and installers, and my own experience, the quality of a carpet pad has everything to do with a first-class flooring job. A good pad will keep carpet from crushing. Traffic patterns will not be so noticeable. An obvious advantage to a great pad is the comfort of walking on it. I can't prove to you that the padding you have installed will add 5 years to the life of a carpet, but I'm convinced that better pads will produce more attractive jobs that last longer.

Vinyl flooring

Most vinyl flooring installed today is sheet vinyl (Fig. 6-3). This type of flooring is desirable because it often can be installed without any seams. Vinyl flooring is nonporous, and impervious to water. It can be cut or burned, but otherwise it is extremely durable and requires minimal maintenance. There are so many colors and patterns available that customers frequently have a difficult time choosing one flooring over another (Fig. 6-4).

It's relatively inexpensive to install sheet vinyl. Compared to tile or wood floors, sheet vinyl is a bargain. It can be installed quickly by just one installer, and the cost of labor and material is much lower than it is for most other types of floor covering. Weighing all of the attributes of vinyl flooring, it's easy to see why vinyl is such a popular floor covering.

Like carpeting, vinyl flooring comes in various grades of quality. The old saying about getting what you pay for often applies to vinyl flooring. A cheap floor will look cheap and won't last long under heavy use. Midrange floors look good and hold up well. Top-of-the-line vinyl, such as the example show in Fig. 6-5, has an outstanding appearance and lasts much longer than the period most people own any one home.

Those familiar with Federal Housing Authority (FHA) loans or Veterans Administration (VA) loans know that materials used in the construction of homes financed with these loans must be approved. Items such as flooring are generally included for approval for FHA and VA loans. I use this as a benchmark when judging quality. This doesn't mean that some flooring that doesn't meet FHA and VA standards is not as good as, or even better than, some approved products. Using

6-3 *Sheet vinyl flooring.* Armstrong World Industries

FHA and VA standards, however, is a fairly safe way to ensure that your customers will not be ripped off by selecting a good-looking but short-lasting flooring. There are so many grades of vinyl flooring that I wouldn't know where to begin explaining them. Just as there is a multitude of grades, the price ranges vary widely. I've seen vinyl advertised for less than $4 per square yard and for more than $35 a yard. Materials that are priced around $15 a yard are typically a good middle-of-the-road material.

Flooring options

When advising your customers on vinyl flooring, there are a lot of little things that should be mentioned. For instance, some no-wax floors are more no-wax than others. A lot of cheaper floors become dull and lifeless. This is not the case with better grades of flooring. I've got vinyl flooring that has never been waxed but looks like it was installed yesterday. On the other hand, I've had some floors installed that looked dull from the day they were put in. This is very important to a customer who wants cosmetic improvements.

The cushion factor in vinyl flooring can be important to some people. Not all vinyl flooring responds the same to footsteps. How

6-4 *A pebble finish in a rubber tile floor.* Azrock Floor Products

6-5 *Elegant vinyl flooring with a border.* Azrock Floor Products

resilient is a particular type of flooring? What is the expected life of the product? In what widths is the flooring available? These types of questions should be asked.

Many kitchens and bathrooms can be covered with a solid sheet of vinyl flooring. This eliminates seams that can come loose, catch dirt, or

Flooring options

just look bad. Some rooms are too large to cover with a single sheet of standard vinyl flooring. You might, however, be able to find a manufacturer who offers vinyl in wider widths. Not all manufacturers offer flooring in equal widths, so if you have a large room to work with, do some homework. If you can find a supplier of wider flooring, it might be worth your while to spend the time looking into it.

Three basic types of sheet vinyl flooring see regular use. The first is *inlaid vinyl*. This product is thick and heavy. Designs in the flooring are inlaid so that they penetrate the thickness of the flooring. If customers are looking for deep colors and sturdy vinyl, this is the type to pick. But you must be prepared to pay more for this heavy grade of vinyl.

Rotovinyls are much thinner and lighter than inlaid vinyls. Designs used with this type of flooring are surface printed and covered with a layer of clear vinyl. Since rotovinyls are made with a core of foamed or expanded vinyl, they are comfortable underfoot and tend to muffle noise.

For customers who want a light, cushioned vinyl that has a lot of give in it, look to *stretch-cushioned vinyls*. This material has enough flexibility to expand and contract as floors shift with the seasons. It's possible for this type of vinyl to hide some imperfections in the subflooring.

While we're on the subject of subflooring and imperfections, let's discuss the importance of subfloor preparation (Fig. 6-6). Contractors engaged in cosmetic improvements typically work with existing conditions and materials. This can lead to a question of whether or not to install new vinyl over old vinyl. This can be done, but I rarely recommend it.

The surface to which vinyl flooring is applied should be flat, level, clean, and free of gaps, cracks, knotholes, and other defects. Even small faults in subflooring can show through a new vinyl floor. Installing new vinyl over old vinyl is, in my opinion, risky. Trying to clean an existing floor covering to a point suitable for a new installation is difficult. You must remove all wax, grease, and dirt. This isn't easy to do.

If an old floor has any humps, bumps, bubbles, or cracks, such imperfections probably will show through the new flooring. It costs a little more to strip out old vinyl for a new installation, but I believe it is well worth the expense, for both the contractor and the customer. Even if the contractor has to absorb the cost of removal, I would suggest doing it. If you install a new floor that comes out looking bad, you are going to lose much more time and money than you would spend preparing a good subfloor for an installation.

Component	Material	Load, psf
Framing (16" oc)	2 × 4 and 2 × 6	2
	2 × 8 and 2 × 10	3
Floor-ceiling	Softwood, per inch	3
	Hardwood, per inch	4
	Plywood, per inch	3
	Concrete, per inch	12
	Stone, per inch	13
	Carpet	0.5
	Drywall, per inch	5

6-6 *Knowing the weight of various building materials can be important when choosing a subfloor, floor structure, and floor covering.*

An alternative to removing existing flooring is to install a thin underlayment over it. The new underlayment provides an ideal base on which to install new vinyl. Increasing the thickness of a floor can create some problems with doorways, so consider this before you commit to a plan. There are plenty of people who will tell you that there is nothing wrong with putting new vinyl down on old vinyl that is tight and in good shape. I can't say it's wrong to do this, but I'd advise you to bid jobs with enough budget to allow for better base preparation.

Even when new underlayment is installed, your flooring installer will need to do some prep work. All seams where the underlayment meets should be sealed. The surface prepared for new vinyl should be flat and smooth. If the cracks between sheets of underlayment are not filled, they will show through most types of vinyl flooring. The same is true for holes created by nails and screws. I've seen vinyl flooring installed in which every row of nails could be seen under the vinyl. This is not what you want. Insist that your flooring installer prep all floors properly.

Don't get caught between a rock and a hard place with flooring installations. You know how painters blame drywall installers and drywall installers blame painters when walls don't turn out well? Guess what? Flooring installers blame carpenters and carpenters blame flooring installers when floor coverings don't look good. Flooring installers will say that the carpenters didn't sink their nails deeply enough or that the wood used was warped. Carpenters will swear that flooring installers didn't prep a job well enough before laying a piece of vinyl. If you have a bad-looking floor, which of these two trades are you going to blame and seek restitution from? When this situation occurs, the contractor is in a difficult spot. But, you can avoid being caught in this trap. Insist that your flooring installer assume responsibility for underlayment, preparation work, and the flooring installation. If you do this, a finger can be pointed at only one person.

It took me a many jobs and quite a bit of lost money to realize that I should limit liability where I could. I started working only with drywall contractors who also offered painting services. My flooring installers had to be willing to take full responsibility for floor prep, or I wouldn't use them. By eliminating the need for the involvement of multiple trades and hands in a job, you reduce the headaches suffered by so many general contractors.

Good flooring contractors can be hard to find. Some flooring installers are interested only in getting in and out quickly. You should spend a little time looking over reference work done by prospective installers before contracting them to do your work. If you buy from a supplier who installs what is sold, you might get piece workers who are paid the same whether a job takes 2 hours or 10 hours. There's nothing wrong with this type of arrangement, as long as the workmanship is good. Some piece workers, however, are not as quality-conscious as they are money-hungry. Find out who your installers will be and check their past performance. Flooring is one part of a cosmetic makeover that remains in clear view after a job is done. This is not a phase of work where you can afford mistakes.

Break the rules

Some of the best contractors earn fantastic reputations when they break the rules. So many houses have carpeting and vinyl flooring in them that it is easy to assume that these are the only two types of flooring to use. Is this true? Of course not.

Wood flooring certainly is a possibility, and so is tile (Figs. 6-7 and 6-8). Wood is expensive and not as popular as it once was, but this doesn't mean that it should never be considered. Take a dining room with stained carpeting as an example. You could replace the carpeting with new material. Doing so would improve the room's appearance, but it would still be a carpeted dining room. Install hardwood flooring, and you've created something special. The cost of installing hardwood in a single room isn't likely to break a customer's budget.

6-7 *Wood flooring used in random widths.* Rutt Cabinetry, Lis King

Wood flooring is available in hard and soft wood. You can buy strip flooring, which often is used in formal settings, or you can opt for plank flooring that provides a cozy country look. Wood tiles also are available. All three of these types of wood flooring have their place. The use of a plank flooring in a family room or kitchen works nicely. Wood tiles give foyers an elegant look. Strip flooring, which is the most common, often is used in living rooms and dining rooms. None of these materials is cheap, but all are affordable in small quantities.

There are many quality vinyl floorings available that simulate real wood so well that from a distance it is difficult to tell the difference between the vinyl and real wood. Product lines are offered in different designs, so you can get strip flooring or squares (Figs. 6-9 to 6-11).

I've always enjoyed installing quarry tile in foyers. There are so many design possibilities that customers just love tile. The fact that tile is extremely durable, easy to clean, and attractive makes it easy to sell. It costs considerably more than vinyl, but when a small area such as a foyer is being done, the cost is not overpowering. And the visual effect is stupendous.

I don't normally recommend tile floors for kitchens in which children are likely to be rambunctious. For one thing, tile gets slippery when it's wet, and tile floors in kitchens tend to get wet and greasy.

Flooring options 67

Another reason why I shy away from tile in a kitchen is the risk of glass breakage. Children, and even adults, sometimes drop glasses or other glass objects. Glass falling on a vinyl floor might or might not break, but you almost can guarantee that it will break if it makes contact with a tile floor. Even though I don't install many tile floors in kitchens, I do use tile for countertops, backsplashes, and walls. There have been times when I've installed tile floors in kitchens at an owner's request. If your customer is not concerned about glass breakage and slippery footing, tile is an excellent flooring choice. By the way, you can get some tile that has built-in traction enhancers to help offset the slippery conditions (Figs. 6-12 and 6-13).

6-8 *A tile floor with color keys.* Azrock Floor Products

Chapter Six

6-9 *Vinyl flooring that simulates stripwood flooring.* _{Azrock Floor Products}

I have installed a lot of tile floors in bathrooms, and have some concern about slippery conditions that can occur. I think this is something that a contractor should discuss with customers. It's hard to beat the beauty of a tile floor, but make sure customers understand that the material can be a difficult one on which to keep their footing.

Slate has been used frequently as a flooring material, especially in foyers. Personally, I prefer tile (Fig. 6-14). Slate gets extremely slippery when wet, and it often turns gray as water dries on it. I don't like either of these characteristics. Plenty of people do like slate, however, and it is another material for you to consider installing.

I have used old bricks for flooring in country kitchens and for hearths around wood stoves and fireplaces. The results have always been good. Old bricks add warmth (in terms of looks) to any type

Flooring options

of rustic setting. One problem with bricks is that they add height to a floor. If you plan to install a brick floor, you must allow for this potential problem. Using half-bricks is one way to reduce the rise in floor level.

There are other ways to get creative with flooring, but wood, tile, vinyl, and carpet are the four most common types of flooring material. Each of the four has a number of subtypes (Fig. 6-15). You and your customers should be able to come up with more than enough ways to generate eye-catching floors within this group of traditional flooring materials.

6-10 *Stripwood vinyl in a kitchen.* Azrock Floor Products

6-11 *Vinyl flooring made to look like wood squares.* Mannington Resilient Floors

Flooring options

Ceramic Tile:
   Sizes range for 1-inch squares to 12-inch squares
   Most tiles have a thickness of $5/16$ inch

Ceramic Mosaic Tile:
   Sizes range from 1-inch squares to 2-inch squares
   Also available in rectangular shapes, generally 1 × 2 inches
   Average thickness is ¼ inch

Quarry Tile:
   Sizes range from 6-inch squares to 8-inch squares
   Rectangular tile is available in 4 × 8-inch size
   Typical thickness is ½ inch

6-12 *Tile sizes.*

Ceramic Tile   Fairly Easy

Ceramic Mosaic Tile   Easy

Quarry Tile   Fairly Difficult

6-13 *Difficulty rating for installing tile.*

6-14 *Ceramic tile in a foyer.* Armstrong World Industries

Flooring options

6-15 *Examples of tile patterns.* Summitville Tiles, Lis King

7

Plumbing propositions

Customers who want to give their homes a new look have plenty of plumbing options to consider. Plumbing might not be one of the first topics that comes to mind when planning cosmetic work, but plumbing fixtures can play a vital role in the appearance of a room. The rooms most affected are, of course, bathrooms and kitchens. Coincidentally, these two rooms are generally considered to be the most profitable rooms in a home to remodel.

A kitchen tends to be the room with the most recapture potential in terms of remodeling dollars. This means that homeowners who invest in kitchen remodeling are more likely to get most, if not all, of their remodeling costs back when they sell their homes. Bathrooms also offer tremendous recapture potential. For homeowners who turn to cosmetic improvements to facilitate quicker sales of homes, kitchens and bathrooms should be at the top of their list of priorities.

Some people love to cook and putter about in their kitchens. Even those who don't enjoy kitchen duty spend a lot of time preparing food, washing dishes, putting groceries away, and doing other kitchen chores. Those who spend a lot of time in a room should strive to be comfortable in it. That's a good reason to dress up a kitchen.

Bathrooms have many uses for homeowners in addition to the obvious ones. Some people enjoy soaking in a hot bath as a form of relaxation. Many parents feel that a bathroom is their only safe haven from the kids. It's the one room in a house where a locked door usually is respected. More than one parent has sneaked off to the bathroom for a short break from parental duties. There are people who spend a lot of time putting on makeup, shaving, and primping for work or a night on the town. Considering the frequency with which a bathroom is used, it makes sense for it to be attractive and comfortable.

Contractors who get involved in kitchen or bathroom remodeling must, in most cases, deal with plumbing. This is not a bad thing, but it requires preparation. Helping a homeowner choose a new style of lavatory can be time-consuming, and picking out just the right faucet can take hours.

Cosmetic improvements can open the door for other improvements. For example, replacing an old sink might be the ideal time to update the piping and valves that serve the sink. Even though this type of work won't show when a job is done, it often is wise to take the opportunity to correct deficiencies during the makeover work. Very little is more frustrating than to finish a nice cosmetic job and then discover a month later that some of the new work has to be torn out to gain access to an old problem.

Kitchen plumbing

Kitchen plumbing can be simple, consisting only of a sink and a faucet. There also might be a dishwasher, a garbage disposer, and additional sinks. Rarely is kitchen plumbing complicated. A kitchen sink and its faucet are the key cosmetic items in terms of plumbing (Fig. 7-1). A grungy old garbage disposer is not seen by guests, and dishwashers count more as appliances than as plumbing. So your main points of interest will be the sink and faucet.

Since you have only a sink and faucet to be concerned with, the cosmetic options for plumbing in a kitchen should be a breeze to understand. Maybe they should be, but don't be fooled. Faucet options alone can be mind-boggling, and the various types of sinks available can keep you looking through catalogs for days.

If you haven't experienced it yet, you will find that homeowners can be very picky when it comes to plumbing. Finding just the right faucet and a sink that offers a unique bowl shape or feature can take several hours. Don't underestimate the importance of plumbing fixtures to homeowners. A corner sink, a three-bowl sink, or a gold faucet might be what the customer wants. Maybe you'll get lucky and have a customer who is interested only in a simple, single-bowl sink with a two-handle faucet. The problem is, you won't know what to expect until you and the customer get involved in fixture selections.

Sinks

An average kitchen sink has two bowls and is made of stainless steel. This run-of-the-mill sink is easy to come by and costs a plumber anywhere from about $40 to more than $100, depending upon the quality

Symptoms	Probable Cause
Faucet drips from spout	* Bad washers or cartridge * Bad faucet seats
Faucet leaks at base of spout	* Bad "O" ring
Faucet will not shut off	* Bad washers or cartridge * Bad faucet seats
Poor water pressure	* Partially closed valve * Clogged aerator * Not enough water pressure * Blockage in the faucet * Partially frozen pipe
No water	* Closed valve * Broken pipe * Frozen pipe
Drains slowly	* Partial obstruction in drain or trap
Will not drain	* Blocked drain or trap
Gurgles as it drains	* Partial drainage blockage * Partial blockage in the vent
Won't hold water	* Bad basket strainer * Bad putty seal on drain
Spray attachment will not spray	* Clogged holes in spray head * Kinked spray hose
Spray attachment will not cut off	* Bad spray head

7-1 *Potential problems and solutions for kitchen sinks.*

of the sink. When looking at stainless steel sinks lined up next to each other, it can be difficult to see any real difference in them. But quality does show through for those who know what to look for. And poor quality stands out quickly to users of the sinks.

Cheap sinks usually are made with a light-gauge metal. Some of the sinks are so flimsy that the installation of a garbage disposal pulls the center of the sink down. I'm not kidding; this really does happen with some sinks. Stronger sinks cost more, but they don't pop and ping like cheap sinks. A sink with soundproofing undercoating is less noisy when water is being run into it. This may not seem important to contractors, but it can be a nice benefit to the user of a sink.

It's not fair to say that all inexpensive sinks are cheap. You can get some very good sinks at reasonable prices, and you can pay a lot for a sink of poor quality. When working with your customers on the selection of a stainless steel sink, look for soundproofing on the bottom of the sinks and give the bowls a push test. Either put your fingers through the drain holes and pull and push, or, with sink displays mounted in counters, push down on the center of the bowl. Do the same to the flat area where the faucet holes are. This will give you a good idea of a sink's sturdiness. If a disposer is to be attached to the sink, go for a little thicker metal.

Not all kitchen sinks are made from stainless steel. Enameled cast-iron sinks are popular in more expensive homes. These sinks are available in vivid colors and various configurations. Make no mistake about it: A cast-iron sink is heavy, and requires a strong counter to support it. Cast-iron sinks don't offer much functional advantage, but do provide more appearance-based options. Since you are doing cosmetic improvements, this may lead you to enameled sinks. Cast-iron sinks are quieter than stainless steel units, but that, in my opinion, is the only practical difference. Any other appeal is cosmetic, and customers will pay dearly for their taste in good looks.

Beyond cast-iron and stainless steel sinks, you get into specialized territory. There are some awesome models from which to choose, but the price tags are scary. Marble-type sinks can be ordered, as can sinks with a variety of other high-grade finishes. Most people, however, will be happy with either a stainless steel or cast-iron sink.

The design of a sink is often as important at the type of material it is made from. Most people want to have at least two sink bowls. Many want a third bowl for salad preparation or a food grinder. If you look through catalogs from plumbing manufacturers, you will see that a three-bowl sink can be had in a number of different configurations. So can two-bowl sinks, for that matter. You will have to spend some time with your customers to determine exactly what style of sink will suit their needs.

It is not unusual for people to want more than one sink in the kitchen. I've done many remodeling jobs where part of the job included the addition of island cabinets and an island sink. These sinks can range from simple, small bar sinks to exotic vegetable sinks. Believe me, there are a lot of possibilities when you are looking for cosmetic improvements in sinks.

Refinishing

Refinishing an existing sink might be worthwhile. This is especially true if the existing sink is old and in keeping with the character of the kitchen. For example, an old-style sink that has an integral backsplash and spigots that come through the back of the sink wall is hard to find and expensive. If your customer has this type of antique sink and wants to retain it, refinishing is the answer to your cosmetic update.

There are companies that will come to a job site and repair or refinish porcelain. This can be done with sinks and bathtubs, and it is a cost-effective alternative to restoration-style replacement. Almost any crack, chip, or discoloration can be corrected by on-site refinishers. You even can have the color of a fixture changed to accommodate a new motif. Let your fingers take a walk through the advertising pages of your local phone directory to find these refinishers.

Replica fixtures

Replica fixtures can be purchased through specialty suppliers. If you want a new sink that looks like an old one, chances are good that someone is marketing such a product. Many companies specialize in replica fixtures. You can locate these sources by reading advertisements in magazines that cater to country kitchens. A little research at your local library will turn up manufacturers and suppliers of specialty fixtures. Check with the person in charge of reference materials at your local library to see what books to check for names, addresses, and phone numbers of such companies.

Faucets

There was a time when faucets were a simple necessity in a kitchen. Now they can be a status symbol or a specialized work tool. There is, for example, a kitchen faucet with a spout that telescopes up to accommodate large pots. This is a functional feature. Single-handle faucets are seen in more new houses than two-handle faucets, but there are people who still prefer to have separate handles for hot and cold water. Some people want a spray attachment on their kitchen faucet, and other people detest these accessories that often leak and cause more trouble than they are worth.

Once you get past basic design features, you have to deal with the finish of the faucet. Will it be chrome or antique brass? A few customers might want a gold-plated faucet. Depending upon the kitchen decor, a faucet that comes in a fashionable color might be the ticket. You can get color-coordinated faucets without much trouble. There are so many faucet styles and features available that even plumbers have trouble keeping up with all of them. Plan on spending considerable time with your customers during the selection of a perfect faucet.

Accessories

There are, of course, accessories available for kitchen sinks. You can get sink-mounted soap dispensers or hot-water makers. If a dishwasher is being added during a remodeling job, you should plan on buying a sink that will accommodate an air gap. These types of accessories affect the number of holes a sink should have. A standard sink has three holes, enough to house a faucet, but no accessories. Some faucets have a built-in housing for a spray attachment, but most don't. Depending on the type of faucet being used, it is not uncommon for a sink to need four holes to hold a faucet with a spray attachment. Start adding soap dispensers, hot-water makers, and air gaps, and you need six or seven holes.

It's not difficult for an experienced plumber to cut additional holes in stainless steel sinks, but when it comes to a cast-iron sink, the possibility of making a decent hole diminishes. Be sure to plan for all the holes that will be needed before ordering a sink. Many plumbers and suppliers will not accept returns on special orders, so if you miscount the number of required holes, you might have an expensive sink that has to go into inventory. Few contractors want to carry this type of inventory, so make sure you know how many holes are needed before you place an order.

Garbage disposers

Garbage disposers don't fit the mold of a typical cosmetic conversion, but they often are added when a kitchen sink is replaced, or when houses that have not been equipped with disposers are upgraded. Old disposers frequently are replaced. It's not difficult to shop for a disposer, but there are a few tips that might be helpful.

A low-horsepower grinder will work, but it is best to install units that run with at least a ½-horsepower motor. They are more expensive, but they usually perform better and are typically worth the extra cost. Another consideration is whether to install a switch-operated disposer or a batch-feed model. Batch-feed models cut on automatically when food waste is pressed into them and cut off automatically when the grinding process is complete. Switch-operated models have to be turned on and off manually.

Batch-feed food grinders cost more than switched models. Is the cost justified? It depends on your customer. Personally, I prefer a switch-operated disposer, but not everyone agrees with me. As long as you know that the two types exist, you and your customer can work out the details of which type to buy.

If you are installing a disposer where one has never been installed before, you should plan on having an electrician run a new circuit for the food grinder. Ask your plumber to check the existing drain pipes to see that they are in good shape and capable of handling the waste from a disposer. Some old houses still have galvanized drain pipes. If this is the case on your job, be advised that the pipes could start clogging up once a disposer is installed. Rust and gunk builds up inside galvanized steel pipes, and although water might drain just fine, adding a food grinder can be too much for old pipes with restricted diameters.

If you find that a sink does have a galvanized drain, don't get too upset. The galvanized pipe usually runs only a short distance to a cast-iron drain. In most cases, the piping is concealed in a wall behind base cabinets. Remove the cabinets, open the wall, and replace the section of galvanized pipe with Schedule-40 PVC pipe. This will reduce the risk of having a completed cosmetic job and a sink that won't drain.

Another word of caution: Check on whether the house where a new disposer is to be installed is served by a septic system. Many jurisdictions prohibit the installation of food grinders on systems supported by private sewage disposal systems. Your plumber should be able to tell you if it is legal and safe to install a disposer, but you can check with the local plumbing inspector if you want to be sure.

Dishwashers

Adding and replacing dishwashers isn't much of a job for a plumber. If a dishwasher is installed adjacent to a sink base, the water supply can be connected to the hot-water pipe serving the kitchen sink. A drainage connection can be put into the piping under the sink to accept the discharge from a dishwasher. Many codes require the use of an air gap when a dishwasher is installed, so keep this in mind when dealing with the number of holes in a kitchen sink. Otherwise, dishwasher replacements and additions are fairly simple.

Bathrooms

Bathrooms typically house most of the plumbing fixtures in a house. It is not uncommon for homeowners to want their bathroom fixtures replaced from time to time. Sinks and toilets are pretty easy to replace, but bathtubs and showers are not so simple. Straight replacements are not the only considerations when beautifying a bathroom. Your customer might want to replace old tile around a bathtub with fiberglass walls. There might be a desire to have old fixtures, such as a clawfoot bathtub, refinished. And there's always the chance that a customer will want a bathroom enlarged and additional fixtures, like a whirlpool tub, installed. There's plenty to think about during the remodeling of a bathroom (Figs. 7-2 and 7-3).

When I think of all the bathrooms that I've remodeled, a few stand out in my mind. An avocado green bathtub, toilet, and wall-hung sink seem to have been the first plumbing fixtures I replaced. If

7-2 *A modern bathroom with a clawfoot tub.* Armstrong World Industries

Plumbing propositions

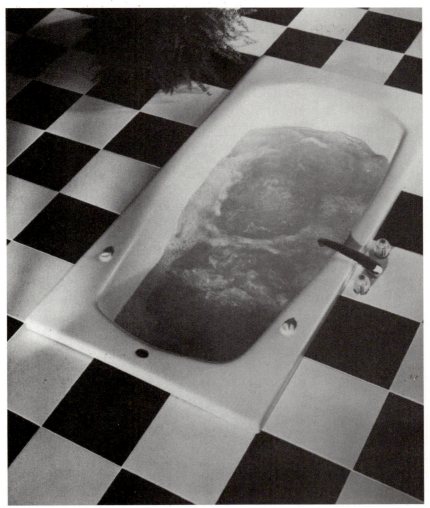

7-3 *A whirlpool tub addition.* HIP

I remember correctly, the floor was white with black specks and was made of some type of composite tile, probably filled with asbestos. The bottom half of the walls was tiled with putrid pink tiles, and the upper half of the walls had layers of ugly wallpaper on it. The toilet sat in a cubicle and a white counter with gold flecks stood beside the lavatory. A rusted metal medicine cabinet was centered over the lavatory and flanked on both sides by fluorescent lights with plastic covers. To be blunt, the room was horrible.

Another bathroom that I remember well had water-resistant paneling that was white with blue dots sprinkled in for color. The plumbing fixtures had been white at one time, but had turned brown with rust stains. A wall-hung sink in the room had individual spigots, both of which dripped constantly. The bathtub had been fitted with an aftermarket shower arrangement that included a chrome pipe extending from the tub filler and a metal rod that supported a shower curtain. Large cracks running through the base of the toilet looked as though someone had worked on them with white caulking time and time again. Water damage had rotted the subflooring, and mildew clung to the ceiling. This, too, was an unpleasant room to encounter.

As I remodeled the bathrooms mentioned above, I was amazed at what a big difference a little cosmetic work could make. The change in the two bathrooms was so magnificent that I decided to devote much of my time to nothing but bathroom remodeling. It was so fulfilling to see dingy dumps turned into sparkling showplaces in such a short time and at fairly affordable costs that I couldn't resist doing more of this type of work. To this day, I still enjoy a good bathroom remodeling job.

Contractors with extensive experience in bathroom remodeling are aware of the challenges that can be encountered. One of the first obstacles to be overcome is the selection of new fixtures. This runs the gamut from style and type to color and price. There are so many possibilities for bathrooms that sorting through them can be a job in itself.

Toilets

Toilets are considered essential equipment in a bathroom. Your plumber can buy a complete toilet set for less than $60, or you can spend several hundred dollars for a one-piece toilet in a designer color. Name-brand, two-piece toilets generally wholesale for between $75 and $100. If you really want to go fancy with a replica toilet, the type with the tank mounted high up on the wall, you can spend well over $500. So, do you recommend spending $50 or $500?

I've done a lot of expensive bathroom remodeling, but I've never found a customer willing to pay for a replica toilet. However, I've had several customers authorize prices in excess of $300 for one-piece, low-profile toilets in high-fashion colors. Most of my customers, however, have been quite satisfied with normal toilets in the $100 range.

Is one toilet better than another? Yes, some are better than others. Depending upon the design of a toilet, one might work better than another. Appearance is also a factor, and some toilets look better than others. For example, many of my past customers have been turned off by toilets with exposed traps. The actual traps are an integral part of the toilet bowl, but the shape of the trap shows through in some models. I've found that models with smooth exteriors are more appealing to customers.

Does a one-piece toilet work better than a two-piece toilet? Not to my knowledge. Because the fixture is all one piece, there is less chance of a leak since the tank doesn't couple to the bowl with tank-to-bowl bolts. Beyond this minor advantage, the remaining attributes of one-piece toilets are all in their appearance, which is quite nice.

Colors can be a key factor in the selection of plumbing fixtures. If you don't already know that manufacturers often have custom colors that can be obtained only with their products, you will discover this to be true as you gain experience with plumbing fixtures. Most manufacturers create their own colors to prompt people to buy their fixtures. This may sound absurd, but it works. I've had customers insist on a particular color, regardless of the fact that a similar, but not exact, color could be purchased from some other manufacturer for considerably less money.

In addition to colors and one-piece or two-piece toilets, you might find a corner toilet to be both interesting, desirable, and helpful when space limitations apply. I installed a corner toilet in the first home I built for myself because I thought it looked neat. Since that time, I've used corner toilets to take advantage of every square inch of bathroom space. Some customers might find a corner toilet disturbing, but others will rave about them.

Style and color are the keys to toilet selection. All toilets perform the same function, but some perform better than others. For instance, a super-water-saver toilet might not flush very well when installed in an old house that has cast-iron pipe with a gentle slope on it. Depending upon local code requirements, you may have no choice but to install a water-saver model. But be advised, some old piping was not designed for such minimal flushes, and you might experience problems with callbacks.

If you are adding a new toilet to a system, you must consider spacing requirements. The plumbing code requires certain amounts of space between fixtures (Figs. 7-4 and 7-5). Workmanship also is a factor in the plumbing code. When a toilet is installed, it must be installed evenly (Figs. 7-6 and 7-7).

Lavatories

Lavatories seem like a simple enough topic, but they can get confusing. For example, you could be dealing with a wall-hung lavatory, a drop-in lavatory (Fig. 7-8), a rim-type lavatory, a pedestal

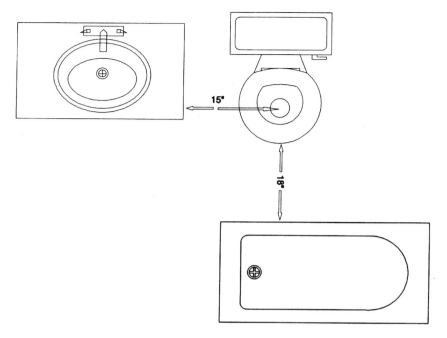

7-4 *Minimum fixture spacing requirements.*

lavatory, or an integral lavatory bowl. Of the five potential choices, which one is best? It depends on the customer's desire and the type of bathroom being cosmetically converted.

Wall-hung lavatories are the least expensive option for a working lavatory, but they are not likely to be chosen when a cosmetic conversion is undertaken. In fact, this is probably the type of lavatory that will be replaced during a conversion. Wall-hung lavatories work fine, but they don't present a prestigious image.

Drop-in lavatories are made to be self-rimming. This simply means that a hole is cut in a counter and the lavatory is placed in it without a rim. This type of lavatory looks good and works well. Lightweight versions, such as those made of plastic, are not as popular as heavy models made of china or cast-iron.

Rim-type lavatories resemble drop-in lavatories, except that they depend on a metal rim to hold them in place. Plumbers generally despise this type of lavatory, because it can be a real pain to install. Customers often complain about rimmed lavatories because dirt collects around the rim and is hard to remove. Based on the variety of choices available, I can't see why a customer or a contractor would opt for a rimmed lavatory.

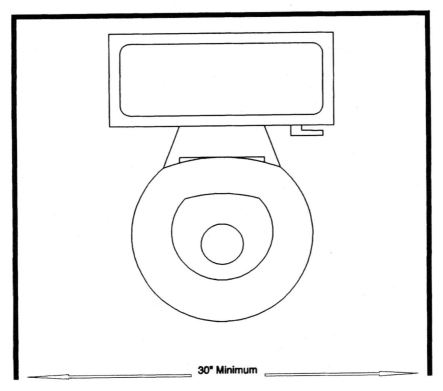

7-5 *Minimum width requirements for a toilet.*

Pedestal lavatories often are considered a luxury, yet some people have no use for them. The downside to a pedestal lavatory is that it doesn't offer any type of storage for linens or toiletries, as a vanity would. The elegance of a pedestal lavatory, however, is hard to beat. And, since this type of lavatory is streamlined, it gives a sleek appearance to a bathroom.

I've installed dozens, if not hundreds, of pedestal lavatories in homes, and I'm familiar with the accolades they get. Top-name pedestal lavatories can wholesale in excess of $300, a lot of money for a lavatory. But I've seen good pedestal lavatories wholesale for less than $75. Shopping for the right product can pay off. Another good thing about pedestal lavatories is that they don't take up a lot of room. This can be a big advantage when working in a small bathroom, especially if the space allotted for an existing wall-hung lavatory is limited.

Integral tops, like cultured-marble tops with an integral lavatory bowl, are my personal favorites. These tops sit on vanities and offer

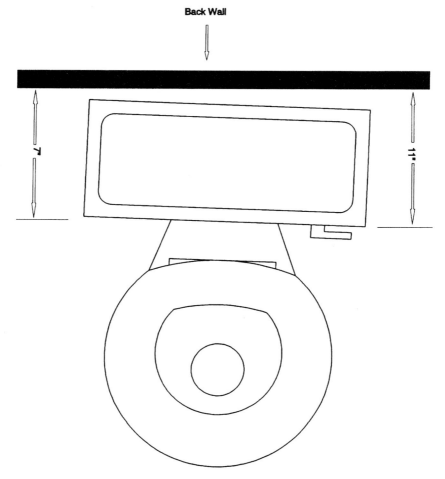

7-6 *Improper toilet alignment.*

the maximum amount of versatility with a minimum amount of risk. Since the lavatory bowl is a molded part of the counter, leaks are nearly nonexistent. Prices for this type of top vary with the size of the top, but they can cost less than $50 and rarely cost more than $175. Of course, there is the cost of the vanity to be considered, and this can easily double the cost of the top. A typical vanity and top might wholesale for around $200, or more.

As a bathroom remodeler, I like the elegance of pedestal lavatories, but I love the functional value of a vanity and top. You are sure to have your own opinions, and so will your customers. My purpose

Plumbing propositions

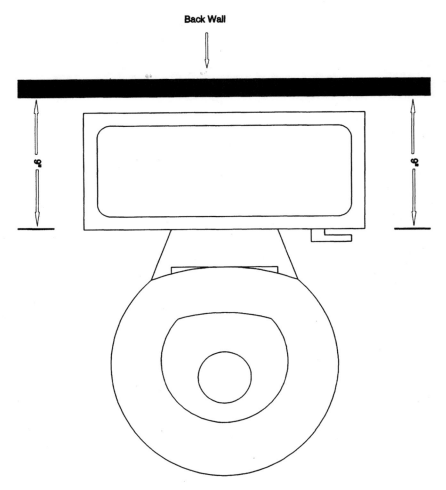

7-7 *Proper toilet alignment.*

is to make you aware of various options to offer your customers, but my money is on the vanity-and-top combination in most cases.

One nice thing about molded vanity tops is that you can order the lavatory bowl in a center position or offset to either side. You also can order a top that contains two lavatory bowls for busy couples who have to prepare for work simultaneously. I've had great success in selling tops with offset bowls that allow for a makeup area at one end and the lavatory at the other. You will find your own niche as you progress in bathroom work, so just offer your customer all of the options and see what works best for you.

7-8 *Drop-in lavatory.* Wilsonart

Lavatory faucets

Lavatory faucets are available in a wide array of styles, colors, and types. Most customers choose either chrome or antique brass for their lavatories, but some prefer polished brass, or even gold. Since moving from Northern Virginia, I haven't had a call for gold faucets in about 10 years. The last two sets I sold wholesaled for around $2500. This is a lot of money for a lavatory faucet, and not many homeowners will invest this type of money in a bathroom. Homeowners often want something special during a cosmetic conversion, so don't be surprised if a customer asks you to quote expensive faucets.

There are two basic types of lavatory faucets: *two-handle faucets* and *single-handle faucets* (Figs. 7-9 and 7-10). Within these two groups, there are a number of different possibilities. I've seen two-handle faucets retail for less than $10 in hardware stores. Similar but better-grade faucets can retail in excess of $100. Based on my experience, a name-brand lavatory faucet will average out at about $85. Many homeowners pay much more for faucets.

When you buy a lavatory, you have a choice between a 4-inch center and an 8-inch center. The 8-inch center is considerably more expensive, and you will find this type of faucet needed most often for pedestal lavatories. Standard lavatories take a 4-inch-center faucet. With prices ranging from under $10 to well over $2500, how will your customers make their decisions? They may need some help from you.

Plumbing propositions

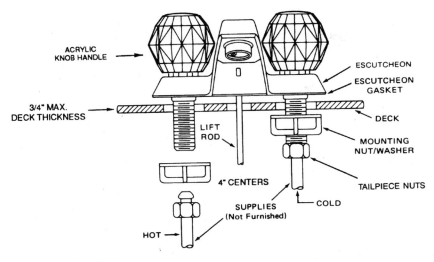

7-9 *Two-handle lavatory faucet.* Moen

Over the years, I've grown partial to a particular single-handle lavatory faucet. It wholesales for about $60 and gives years of dependable service. If anything goes wrong, a single cartridge can be replaced to rebuild the entire faucet. Another brand I've had good

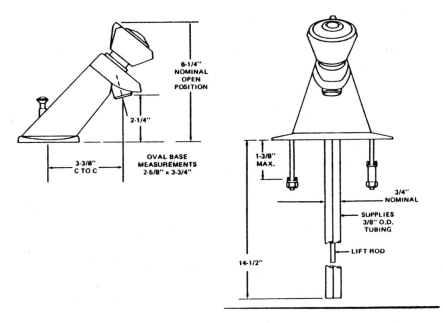

7-10 *Single-handle lavatory faucet.* Moen

luck with uses a ball assembly, springs, and a lot of O-rings. This type of unit is good, but it's more trouble to work on than my favorite. I'm sure your plumber has a preferred faucet, and you should inquire as to what that faucet is.

I can't tell you what faucet is best or which faucet provides the most bang for your buck. My experience is only that, my experience. You have to work through situations on your own and decide what is best. Talk to your plumbing contractor and get some recommendations. Most faucets, even the $10 faucets, do a fair job and don't create a lot of problems. If you have your plumber supply you with brochures, you can show them to your customers and let them make their own decisions. Remember though, 8-inch-center faucets tend to be pricey.

Bathing units

The bathing units in bathrooms are the most difficult fixtures to replace. This is true of both bathtubs and showers. If you can refinish an existing bathing unit, your customer will save money. And, there should be no reason why this can't be done. Many companies offer in-house restorations of existing bathing units. Tile showers and metal showers are exceptions to this concept, but bathtubs, whether they are fiberglass, steel, or cast iron, can be repaired, retouched, refinished, and reincarnated. You even can change the color of the fixture. A talented refinisher can give a fiberglass shower a major makeover.

When customers want to hold their remodeling to a minimum, you should consider having existing bathing units refinished. To pull these units out and replace them is not only expensive, it is disruptive. A toilet can be replaced in less than an hour. Lavatories can be replaced in less than two hours. Neither of these jobs affect the existing walls in a bathroom. To remove and replace an existing bathing unit takes several hours of plumbing time and a lot more in wall repair and painting. If you can have a refinisher come in and do a facelift on a tub or shower in a couple of hours, your customer is way ahead in time and money.

It's not always the finish or color of a bathing unit that cries out for cosmetic help. Some bathtubs don't have shower surrounds. Modern bathtubs are typically tub-shower combinations, but this has not always been the case. Ceramic tile used as a shower surround can leak. The desire to add or replace a shower surround might prompt a homeowner to make a cosmetic change. It is usually not feasible to get a one-piece, tub-shower combination unit into an existing home.

These fantastic bathing units typically are limited to new construction. But there are options available that give similar results in a cost-effective manner.

Achieving the benefits of a one-piece tub-shower combination in an existing home can be done in many ways. Some are much better than others. For example, I would stay away from bathroom paneling that is supposed to imitate tile or similar substances. I've used some of the 4 × 8 sheets of bathroom wallboard during my career, and I've never had a pleasant experience with it. For my money, this material is a waste of time and reputation.

There are a few types of tub surrounds that deserve respect. Both are manufactured by reputable companies and do as good a job surrounding a tub as can be expected of any material. Ceramic tile makes a good surround, but the grout joints can discolor and be hard to clean. As the joints age, they can develop leaks. This won't happen with the types of surrounds I suggest you use.

If you pick up a Sunday paper and browse the advertisements, you are likely to see some super-cheap deals on tub surrounds. I've seen them for less than $40. Are these units any good? I can't say, because I've never used them. The tub surrounds I use wholesale out at just under $300, which is quite a bit more money. But I've never had one leak or fall off a wall, something I've heard rumors about cheaper versions doing. I hate callbacks, and I don't like the idea of getting a reputation for being a fly-by-night contractor. This is why I insist on quality materials. If potential customers aren't willing to pay for professional-grade materials, I won't take their jobs. Maybe I'm old-fashioned, but I believe that if a job is worth doing, it's worth doing right.

I've been remodeling houses for most of my middle-aged life, and I feel this gives me a little edge over rookies who think they can create a booming business on low prices alone. Price is important, but it should never be looked upon as being more important than the quality a customer receives. My $300 tub surrounds look wonderful. They don't leak, and I'm convinced they won't cave in on my customers during a shower. If I lose a job to a contractor who is quoting a $39.95 tub surround, I believe I'm better off not to have the job in the first place.

Some contractors will gladly apply a tub surround over old, leaking tile walls. I won't do this. If a tile wall is leaking, I insist on having the tile and drywall removed before a repair is made. It's important to me to install water-resistant drywall and to provide a solid, dry, clean base for my tub walls to adhere to. This requirement costs me some jobs, but I'm willing to lose work that must be done at less than professional standards.

There is absolutely nothing wrong with refinishing a tub or shower. If an existing bathing unit is in good shape, you are doing your customer a favor to provide refinishing services. There are some occasions when existing walls are sound enough to install tub surrounds without replacing them. You or your plumber should be competent to decide when this is the case. Don't fall into the bidding-war syndrome in which you are willing to sacrifice quality for work. You may win the bid, but your reputation can suffer. If you are in business for the long haul, it's important that you maintain a quality of workmanship that is unsurpassed. Stooping to the level of some hungry contractors, little more than unemployed workers who placed an ad in the local paper, will not propel your business into a bright future.

If you are going to install tub surrounds, do it right. Use only the best tub walls available, and make sure the wall preparation is in compliance with the manufacturer's recommendations. Do this, and you will have happy customers and a bright future.

Bathroom remodeling can be complicated. Removing existing bathing units and replacing them with sectional tub-shower units requires a lot of work with walls and wall coverings. Keep this in mind when you bid a job. Refinishing and adding tub walls will not always meet your customer's expectations. There will be times when you have to physically remove bathing units and replace them. This type of work gets involved.

If you are asked to replace an existing bathtub with a whirlpool, you might have to create more space. Some whirlpool tubs will fit in the space of a standard tub, but these units often are too small to suit customers. Plumbing is a phase of cosmetic remodeling that can require a lot of knowledge and patience. Later on, you will be given some specific advice on bathroom makeovers. For now, we are going to move on to the next chapter and discuss electrical options.

8

Electrical options

Electrical options are often not given the credit they deserve by people planning cosmetic conversions. They tend to take light fixtures for granted, and rarely think about switch and outlet covers. All of these components of a home contribute to its overall appearance. Even outlets and switches can play a part in how a house looks.

Have you paid any attention to the switch and outlet covers in your home? Probably not; most people don't. If you were visiting someone else's house, however, you might make a mental note of how dirty their box covers were or notice that the cover plate on an outlet was broken. It's surprising how little things can catch a person's attention.

Light fixtures can date a house. An experienced contractor can walk into a house and make a good guess about its age just by looking at the light fixtures, assuming that the fixtures are original issue. Yet many homeowners never consider having new fixtures installed, unless they are doing a complete remodeling job. There is no reason why homeowners couldn't, or shouldn't, upgrade their electrical fixtures periodically.

As a general contractor, you should be aware of the importance of light fixtures to the appearance of a home. If you've been doing remodeling work for long, you probably know that some light fixtures produce different effects. Ceiling fans, chandeliers, track lighting, and other types of light fixtures all can enhance a room.

I once remodeled a house in which the light in a main hallway was just a plain, while globe. The hall was dark and uninviting. During the remodeling process, I had my electrician install a second light in the hall. In addition to adding a light, we changed the type of light fixture to produce a new look. The fixtures chosen were brass with cut-glass inserts. Even when the lights weren't on, they looked nice. When they were on, they cast mesmerizing shadows on the freshly painted walls. It really doesn't take much to make noticeable differences when dealing with electrical fixtures.

Switch covers

Switch covers and switches come into contact with human hands frequently. Dirt, grease, and oil are left on switch covers by these contacts. As time passes, the dirt builds up and becomes more noticeable. Residents of a home don't usually perceive a difference in the color of their switches and switch covers. But if a new cover were to be held next to one that had been in place for a couple of years, the difference would be obvious.

Switch covers are cheap and easy to replace. There's no excuse for painting a room and leaving old switch covers in place. Replace them with new ones. Consider using something other than the standard plastic covers that usually come in an ivory or brown color. Go with switch covers made of wood or brass. If you are redecorating a child's room, get a switch cover in the shape of a clown or some other figure. Dress up freshly painted walls with eye-catching switch covers.

Light switches don't usually discolor to a point where replacement is necessary. There may be occasions, however, when it is advisable to have your electrician replace a switch. For example, a room with brown switches might look better with ivory switches. In some cases it might be impressive to replace standard switches with dimmer switches, a move that can prove very effective with chandeliers and mood lighting.

Outlets

Receptacles and their covers should be treated like switches and their covers. When dressing up a room, don't ignore outlet covers. Plastic covers sometimes crack, and often discolor with age. Get rid of ugly covers and replace them with covers that will add to a room. Again, you can use designer covers or new plastic ones, but don't reinstall the old ones after painting a room.

Ceiling Fans

Ceiling fans once were used to cool buildings on hot days (Fig. 8-1). They are still used for this purpose, but they also are installed as decorative objects. They can be installed on vaulted ceilings and flat ceilings. Light kits can be purchased with ceiling fans to give them a dual purpose. My house has three ceiling fans with light kits installed. There is one in the master bedroom, one in the sun room, and another in the living room. All of my fans see a lot of use.

Installing a ceiling fan in a room offers several advantages. If the fan has a light kit, it brightens the room considerably. A reversible fan

Electrical options

8-1 *Ceiling fan, without light kit.* Nutone

can help reduce heating and cooling costs. And a fan adds flavor to a room. The selection of fans available is large, and prices range from very little to a lot. A good fan can be bought for under $50, but some decorative fans cost much more.

It is sometimes possible to replace a standard light fixture with a fan, but be sure the electrical box is strong enough to support the weight and motion of a fan. Ideally, fans should be installed on boxes that were intended to hold the heavy weight of a fan. Depending upon what your electrician charges for labor, you could get a ceiling fan installed for less than $100. This is certainly an affordable way to change the look of a room.

Lights

Lights are the chief element of an electrical cosmetic conversion. You can replace lights and add lights to change the complexion of a room. The price of fancy light fixtures can shock your customers, so be selective in what you show them and how you present the options. I'm not suggesting that you steer customers away from expensive fixtures, but it is a good idea to prepare them for the high prices they might see.

Before your customers go to a lighting supplier to look at fixtures, discuss with them how their lighting allowance is figured. Did you base the allowance given to them on wholesale prices or retail prices? It's not unusual for light fixtures to carry a huge markup, so it's important that your customers know how to track their lighting budget.

I normally figure my lighting allowances on wholesale prices. On one occasion I forgot to tell my customers that their budget was based on contractor's prices and not on sticker prices. The customers went to a showroom with their $500 lighting allowance (this was many years ago), and found that they couldn't afford to buy more than one chandelier and a few lights. The chandelier cost more than half of their total budget. Frustrated, the customers called my office in something of a panic. I went to the showroom and explained to them how to read the discount codes. By doing that, and explaining that they were given an allowance based on wholesale prices, I was able to calm the customers. I learned right then and there always to make a point of explaining to my customers, before they go shopping, how their lighting allowance works.

You don't have to engage in full-scale remodeling to make astounding differences with light fixtures. Simply replacing old fixtures with new ones can work wonders for a room. Consider the earlier example about hall lighting. I had the electrician remove what was probably a $15 light and replace it with one that cost about $35. The result was worth hundreds of dollars, but cost very little.

Many houses, especially those built in volume by tract builders, have cheap light fixtures. Bedrooms that are equipped with ceiling lights frequently have what I call bug-catcher lights. These are the lights with little glass dishes hiding the light bulbs. Some of them have round shades and many have square shades. Sometimes they have a simple design, like wheat, on the shade. This type of lighting is functional, but it doesn't do much for the charm of a room.

What types of lights are generally found in hallways? The most common type I seem to find has either a black or brass base with a white ball-shaped globe. Again, the fixture is functional, but that's about all it is. There are many attractive types of light fixtures available, but many houses are equipped with the most basic and least expensive lights available.

What types of lights are common in bathrooms? Depending upon the age of a home, many different styles of lighting can be found in bathrooms. There is usually some type of ceiling light in old bathrooms, and of course, there is the ugly fluorescent fixture either above or on both sides of the medicine cabinet. Newer bathrooms might have strip lighting in the mirror area and are likely to have a fan-light combination in the ceiling. Whatever is there usually can be improved upon.

Kitchens often have some type of round ceiling light in the center of the room with a second light, often a fluorescent tube, over the kitchen sink. Rarely is there any under-cabinet lighting, which makes a world of difference in a kitchen. Too many kitchens are dark from inadequate lighting, so this is one room you usually can help immensely by adding new lights or by replacing old lights with more efficient ones. Consider installing recessed lighting in kitchens (Figs. 8-2 and 8-3). Any additional light that you can provide should be advantageous.

Dining rooms generally have a traditional chandelier. Some of these are fine, and others look like they were a prize in a box of caramel corn. Who puts a $40 chandelier in a formal dining room? Would you install wainscotting, wallpaper, chair rail, crown molding, and new flooring, while leaving an old, cheap chandelier hanging in the middle of a room? I hope not. Dining rooms deserve something a little better than a bargain-basement light fixture.

Family rooms are notoriously dark. This is especially true of family rooms that have been created in the basements of homes. Builders and remodelers trying to save a few bucks don't bother with ceiling lights. They install a switched outlet and go on about their business. I'm a firm believer in overhead lighting or some type of wall lighting. The idea of lighting a room with lamps alone doesn't appeal to me. Most family rooms are good candidates for track lighting (Figs. 8-4 and 8-5).

We could continue a room-by-room checklist, but specific rooms will be covered in subsequent chapters, so let's talk in more general terms. You've seen several examples of where and when new light fixtures can make a big difference. Up to now, the discussion has

8-2 *Recessed lighting in a kitchen.* Plain & Fancy

100 Chapter Eight

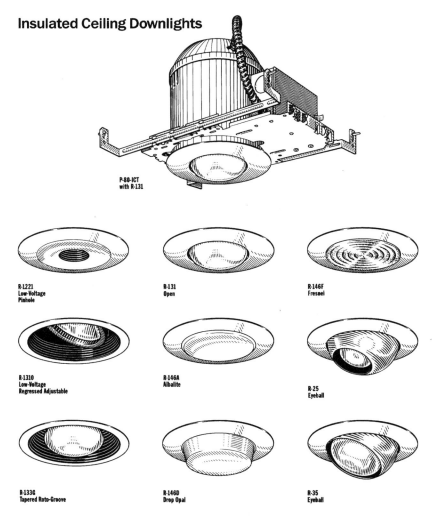

8-3 *Types of recessed lighting.* Nutone

been limited to the interior of a home. Exterior lighting is also worth discussing.

Exterior lighting

Exterior lighting for most houses consists of some type of wall-mounted light by each door and maybe a couple of spotlights or floodlights thrown in for good measure. This isn't a bad start, but it might not be enough. Before we start talking about adding lighting, let's discuss upgrading what already exists.

Low-Profile Track

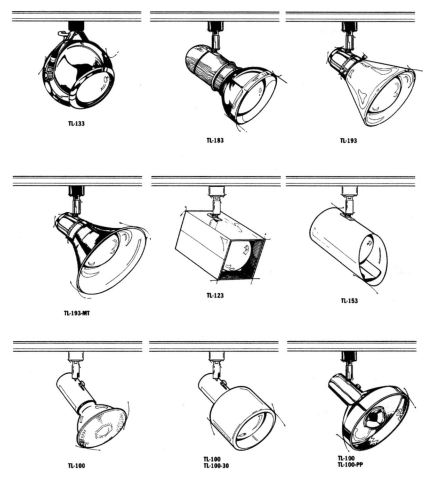

8-4 *Track lighting.* Nutone

I've been building and remodeling homes for a long time, and it amazes me how little some things change. Exterior lighting is an example. For as long as I can remember, a typical door light has had a black base and either a canister-type or ball-type globe. The new lights I see other builders installing now are not much different from the light on my parents' home some 30 years ago. Today's version is plastic, instead of metal, but that's the only key difference.

Coach-lantern lights have been popular for as long as I can remember. Today, you can buy a plastic version for less than $25. This is a step up from the canister or ball lights, but it doesn't make a

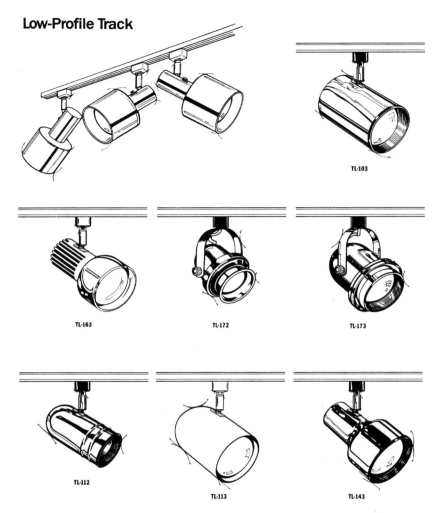

8-5 *Track lighting.* Nutone

powerful statement. Buy a polished brass equivalent and you are on the right track, even though it will cost considerably more. Depending upon the house style, it doesn't hurt to break away from tradition and get creative. One word of caution: Check covenants and restrictions before you replace exterior light fixtures. You might find that certain finishes or styles are prohibited.

Spotlights and *floodlights* are fairly standardized in form and function. Unless their housings have discolored or rusted, there's not much point in replacing them. One exception might be the replacement of old light holders with new, motion-detector-style housings.

Otherwise, I can't see making a big deal about replacing this type of exterior lighting.

Post lanterns were in vogue when I was growing up, but I don't see many of them installed at houses being built today. I'm not sure why this is, but I assume it's a matter of cutting a few hundred dollars off the price of a house. Post lanterns do help to illuminate the front of a home, the driveway, and the walkway. They are practical, and not all that expensive. You might want to consider offering your customers an option of having a post lantern installed.

Walkway lights have all but replaced post lanterns in my part of the country. The little lights come in a variety of sizes and shapes. The ones at my house are made of black plastic and sit low to the ground. They illuminate my walkway very well, and they are attractive to look at from the road in front of my house. My walkway lights are on a sensor that cuts them on and off depending upon the degree of daylight available. It seems to me that my wife paid less than $80 for the entire setup, and I installed it in less than two hours. The effect of the lights on the front of our house was terrific. This is a cost-effective way to brighten up the lawns of your customers.

Accent lighting isn't common in residential settings, but it often is used for commercial buildings. I'm talking about lights, placed in front of a building, that shine on the exterior of the building. They usually are white lights, but colored lights are sometimes used for special effects. If your customer's home has a particular feature that deserves attention, this is one way to make sure it gets it. For example, a large porch with marble columns and stairs could be the target for accent lights.

Garden lighting is not common for houses, but it's very effective under the right circumstances. If your customer has landscaping that cries out for attention, some garden lighting will get it. You might spotlight a fountain in a fish pool or create a pattern in a rock garden. The lights will provide interesting visual appeal while showing off the garden features at night.

Security lighting is more popular than ever. It's sad that our society has come to a point where people don't feel safe unless they can light up their property like a prison yard. Security lights can be rigged with motion detectors or wired to come on only when switches are thrown. The lights can be mounted on posts, in trees, or on parts of a home. This may have little to do with the cosmetics of a house, but it is a service to consider offering your customers while you have an electrician on the job.

Since electrical work normally requires the services of a licensed electrician, you probably won't be doing a lot of it yourself. However, it could be helpful to familiarize yourself with the electrical code. For example, do you know the amperage needs for different household appliances (Fig. 8-6)? Have you ever checked electrical boxes to determine whether they contained too many wires? You can depend exclusively on electrical inspectors, but it never hurts to be knowledgeable about what the various trades are doing (Figs. 8-7 to 8-9).

Appliance	Amps	Volts
Clothes Dryer	30	120/240
Clothes Washer	20	120
Dishwasher	20	120
Kitchen Range	50	120/240
Water Heater	30	240

8-6 *Amperage needs for household appliances.*

Box Shape	Outside Dimension, in Inches	Wire Size, #6	Wire Size, #8	Wire Size, #10	Wire Size, #12	Wire Size, #14
Square	4 × 1¼		6	7	8	9
	4 × 1½	4	7	8	9	10
	4 × 2⅛	6	10	12	13	15
	4¹¹⁄₁₆ × 1¼	5	8	10	11	12
	4¹¹⁄₁₆ × 1½	5	9	11	13	14

8-7 *Number of conductors allowed in a square electrical box.*

Box Shape	Outside Dimension, in Inches	Wire Size, #6	Wire Size, #8	Wire Size, #10	Wire Size, #12	Wire Size, #14
Round and Octagonal	4 × 1¼		4	5	5	6
	4 × 1½		5	6	6	7
	4 × 2⅛	4	7	8	9	10

8-8 *Number of conductors allowed in a round or octagonal electrical box.*

Box Shape	Outside Dimension, in Inches	Wire Size, #6	Wire Size, #8	Wire Size, #10	Wire Size, #12	Wire Size, #14
Rectangular	2 × 3 × 2¼		3	4	4	5
	2 × 3 × 2½		4	5	5	6
	2 × 3 × 2¾		4	5	6	7
	2 × 3 × 3½	3	6	7	8	9

8-9 *Number of conductors allowed in a rectangular electrical box.*

9
HVAC modifications

HVAC modifications are probably one of the lowest considerations on the list of a homeowner contemplating cosmetic remodeling. To some extent, there is nothing wrong with this. Most heating and cooling systems have very little to do with the cosmetic value of a home. There are, however, a few occasions when a heating or cooling system should be considered for cosmetic work.

Think of any situation you can imagine in which a heating or cooling system might have a detrimental effect on the cosmetic appeal of a home. Go ahead. Think. See what you can come up with. Just stop reading for a moment, close your eyes, and envision jobs where elements of the heating or cooling systems are in the way or ugly.

Okay, how many situations did you come up with? Did you note that radiators can be a distraction and an obstacle for furniture placement? Well, they can be and this might be a situation in which modifications would be in order. I suspect you thought of exposed ductwork hanging from the ceiling of a basement that has been finished into living space. This, too, is a situation where some cosmetic work would be in order.

How about baseboard heating units that have become dented and rusted? Registers and grills used with forced-air systems might suffer from discoloration or rust, a cause for cosmetic action. Here's one you might not have thought about. The outside air handler for a heat pump that is placed in a prominent location, detracting from a home's exterior beauty. Have I convinced you yet that there really are times when heating and cooling systems figure into a cosmetic conversion? I hope so, because they do.

How many times have you walked into an old house and seen steam pipes running from the floor to the ceiling? Enclosing the pipes would be helpful in cosmetic terms, but a lot of the pipes are painted and ignored. Is this right? It makes it easier to access the pipes if there is a problem, but it doesn't remove the eyesore or the risk of someone getting burned on the hot pipes. This is yet another piece of proof that heating and cooling systems can require the attention of a cosmetic contractor.

Exposed ductwork

Any contractor who has been in the business for any length of time has seen homes in which exposed ductwork encroached on living space. This is usually the case in basement living space, but it sometimes occurs in other areas. Why do remodelers finish off a room and leave ductwork as a ceiling ornament? I don't know, but some of them do.

It is not always feasible to install a dropped ceiling to conceal ductwork, although this frequently works. Sometimes, due to low headroom, ductwork has to be boxed to be hidden. When this is done, it can be just as distracting as the ductwork was. A contractor who boxes in ductwork should do it in a such a way that the chase box has more of a purpose than just camouflaging ductwork. For example, you could install recessed lighting in the box to give it more personality and purpose. Track lighting could be attached to the box, or trim might outline the box to create a picture-frame effect, enabling the homeowner to use the box as a showcase for photos, awards, or other collectibles. The point is, don't just box in ductwork and leave the box hanging there without any other purpose.

It might help your cause to change the type of ductwork being used. Going from round ductwork to rectangular material might give you the space you need. The opposite could also be true. Before making this type of decision, talk to your HVAC experts for advice on what size limitations apply (Figs. 9-1 to 9-4).

Exposed piping

Exposed piping is common in older homes. This is true of both plumbing and heating pipes. It's very common to walk into the main hall of an older house and see two pipes running between the first and second floors. The pipes usually are in a corner, and often are painted in an effort to have them blend in with the walls. Many people don't find these pipes offensive, but some do. I have a special cosmetic camouflage for this type of situation.

Cubic Feet per Minute (CFM)	Round Duct Size, in Inches	Rectangular Duct Size, in Inches
50	4	4 × 4
75	5	4 × 5 & 4 × 6
100	6	4 × 8 & 5 × 6
125	6	4 × 8, 5 × 6, & 6 × 6
150	7	4 × 10, 5 × 8, & 6 × 6
175	7	5 × 10, 6 × 8, 4 × 14, & 7 × 7
200	8	5 × 10, 6 × 8, 4 × 14, & 7 × 7
225	8	5 × 12, 7 × 8, & 6 × 10

9-1 *Conversion table for 4-inch to 8-inch round branch ducts.*

Assume that you are working with an old two-story home. The front door opens into a hall, and from there you have a choice of going up stairs, down the hall, or through doors on either side. This is a common layout in older homes. As you walk in the front door, there are two steam pipes running from floor to ceiling. They are near a corner, but not quite in it. The owner of this house wants you to do something to get rid of the pipes, or to at least hide them in an attractive manner. What are you going to do?

Relocating the pipes can be expensive, so hiding them is usually the quickest, easiest, and least expensive solution to the problem.

Cubic Feet per Minute (CFM)	Round Duct Size, in Inches	Rectangular Duct Size, in Inches
250	9	6 × 10, 8 × 8, & 4 × 16
275	9	4 × 20, 8 × 8, 7 × 10, 5 × 15, & 6 × 12
300	10	6 × 14, 8 × 10, & 7 × 12
350	10	5 × 20, 6 × 16, & 9 × 10
400	12	6 × 18, 10 × 10, & 9 × 12
450	12	6 × 20, 8 × 14, 9 × 12, & 10 × 11

9-2 *Conversion table for 9-inch to 12-inch round branch ducts.*

Cubic Feet per Minute (CFM)	Round Duct Size, in Inches	Rectangular Duct Size, in Inches
400	10	4 × 20, 7 × 10, 6 × 12, & 8 × 9
450	10	5 × 20, 6 × 16, 9 × 10, & 8 × 12
500	10	10 × 10, 6 × 18, 8 × 12, & 7 × 14
600	12	6 × 20, 7 × 18, 8 × 16, & 10 × 12
800	12	8 × 18, 9 × 15, 10 × 14, & 12 × 12
1000	14	10 × 18, 12 × 14, & 8 × 24

9-3 *Conversion table for 10-inch to 14-inch round main or trunk ducts.*

How do you hide the pipes creatively? That's easy. There are two good ways of hiding the pipes with minimal cost.

The first, and least expensive, way of concealing the pipes doesn't involve much. You will need two finished trim boards. Nail some strips of 2 × 4 material to the walls, floor, and ceiling to provide a nailing surface. Then install the two trim boards to create a box. Since the pipes are close to a corner, you need only the two boards to make a complete enclosure. The trim boards can be screwed to the

Cubic Feet per Minute (CFM)	Round Duct Size, in Inches	Rectangular Duct Size, in Inches
1200	16	10 × 20, 12 × 18, & 14 × 15
1400	16	10 × 25, 12 × 20, 14 × 18, & 15 × 16
1600	18	10 × 30, 15 × 18, & 14 × 20
1800	20	10 × 35, 15 × 20, 16 × 19, 12 × 30, & 14 × 25
2000	20	10 × 40, 12 × 30, 15 × 25, & 18 × 20

9-4 *Conversion table for 16-inch to 20-inch round main or trunk ducts.*

nailers, so that easy access to the pipes is available if repairs are ever needed. Paint or stain the trim boards, and you can be done. I'd add a few brass coat hooks to make the box more useful, but that is only a suggestion.

The foregoing method is the one I used when I first started hiding vertical pipes. Later, I refined the process and came up with a more extensive method that serves multiple purposes. Go ahead and create the first box, as described in the first example. Then build a second box some distance down the wall from the first one. You now have the equivalent of two columns. Your next step is to fill in the middle. You can do this with a coat rack, an umbrella stand, a bench seat with a hinged top, or a mirror, among other things. When you're done, the completed job makes a very handy hallway organizer for people coming and going. Creative ideas such as this one help cosmetic contractors make a name for themselves.

To avoid boxes, it is usually possible to move exposed pipes and conceal them in existing walls. This type of work, however, can get expensive. Walls normally have to be opened up and later sealed and finished. The pipe relocation can result in problems, such as old piping breaking or starting to leak at joints. Once old pipes are messed with, they can give a contractor all sorts of trouble. Your plumbing or heating contractor can most likely move the pipes, but you will reduce risks and probably save money if you can convince your customers to go with some type of box arrangement.

Bad baseboard units

Bad baseboard units can make a house look old and abused. It's not unusual for baseboard heat covers to rust. You could remove the covers, sand them, and repaint them. If you do this, use paint that is intended for use on heating covers. The intense heat will cause problems with regular paint.

If you don't want to bother with sanding and painting, you can buy new covers for the heating system. Covers, both front and back, can be replaced without disconnecting the heating pipes and elements. This makes a cosmetic upgrade simple and cost-effective.

Radiators

Radiators still can be found in many older homes. The size of radiators varies considerably, both in width and height, but all of them are large enough to present some problems with furniture placement. Households with small children run some risk when radiators provide heat. If children touch the heating units, they could be severely burned. People who have lived in homes where radiators were used are familiar with the noises associated with them. Residents get used to the noise, but it can still be annoying, especially to guests.

Builders no longer install radiators in new houses. Many remodelers find themselves replacing radiators with newer types of heating systems. Some homeowners complain about the floor space they loose to the big cast-iron monsters that produce radiant heat. It's not unusual for radiators that are moved during remodeling and then re-installed to leak. All things considered, it is sometimes best to eliminate radiators. There is a strong market for used radiators, since they are becoming rare. If you take a job where radiators are to be removed, don't scrap them. Sell them to suppliers or individuals who need them. There's good money in this.

Outside equipment

Outside HVAC equipment can be detrimental to the curb appeal of a house. While there is nothing wrong with having an exposed air handler, the unit and its piping do little to improve the look of a house. Fortunately, these units are normally placed to the rear of homes, so they are not visible from the front yard. But if customers use their back lawns or have gardens in back of their homes, they might not want to look at an air handler every time they venture out back.

It's not normally practical to relocate outside HVAC equipment. And unless the type of heating system is changed, the equipment can't be eliminated. This leaves cosmetic contractors with one option, and that is to hide the equipment. This is easier than you might think.

Air handlers can't be enclosed in a way that would inhibit air circulation. That rules out some options, but there is one simple, inexpensive option that always works and is usually embraced by property owners. Do you know what it is? Lattice is the answer.

You can use lattice to enclose an air handler, because air will flow through the lattice. If your customer wants more concealment, build the lattice screen large enough to enable your customer to grow climbing vines that will fill in the visual voids of the lattice. As long as you don't crowd the air handler with the lattice, air can reach the system by coming in over the top and past the foliage.

Don't lock yourself into building a square box. A pyramid-shaped screen works well. The base of the triangle starts at the home's foundation and the apex is at a point past the air handler, out in the yard. Staggered panels of lattice can be used to screen off outside HVAC equipment. Considering how easy lattice is to work with, your options are nearly unlimited.

New HVAC systems

New HVAC systems usually are not installed for cosmetic reasons, but they could be. The best example is probably the replacement of radiators with a less-intrusive type of heating system. To stay on track with HVAC systems, you have to be aware of the abbreviations used with equipment (Fig. 9-5). Not only do you have to know the language, you have to be prepared to guide your customers along the path to a perfect system.

Btu	British thermal unit
Btuh	BTU per hour
CMF	Cubic feet per minute
D.D.	Degree day
E	Efficiency
ESP	External static pressure
GPM	Gallons per minute
H	Convective heat transfer coefficient
HL	Heating load
In WG	Inches of water gauge pressure
KW	Kilowatt
KWh	Kilowatt hour
MCF	1000 cubic feet
OAT	Outside air temperature
Q	Heat flow
R	Thermal resistance
RAT	Room air temperature
T.D.	Temperature difference
TON	12,000 Btu per hour
V	Volts

9-5 *Symbols and abbreviations used in the heating trade.*

I have replaced radiators with hot-water baseboard heat and with heat pumps, so people do authorize this type of work. Getting into a full HVAC conversion can be a lot of work and can require extensive planning. To begin with, someone has to compute the heat gain and heat loss of the building being served (Figs. 9-6 and 9-7).

Windows

Doors

Outside walls

Partitions between heated and unheated space

Ceilings

Roofs

Uninsulated wood floors between heated and unheated space

Air infiltration through cracks in construction

People in the building

Lights in the building

Appliances and equipment in the building

9-6 *Common sources of heat gain.*

Windows

Doors

Outside walls

Partitions between heated and unheated space

Ceilings

Roofs

Concrete floors

Uninsulated wood floors between heated and unheated space

Air infiltration through cracks in construction

9-7 *Common sources of heat loss.*

Unless you have a lot of experience with HVAC systems and remodeling, don't even talk about possibilities with customers unless you have your HVAC subcontractor with you. There is a multitude of traps that you can step in when making a heating or cooling conversion. For example, adding air conditioning to a house that has only forced hot-air heat can mean installing all new ductwork. This is something a lot of general contractors don't know.

HVAC modifications

When a combination heating and cooling system is designed, the ductwork is sized to meet the needs of air conditioning. This allows both the heating and cooling to work properly. If the ductwork was sized to meet heating requirements, results of the air conditioning might not be acceptable. I've known remodelers who thought they could add air conditioning to an existing duct system for a heating unit. This shouldn't be done. I'm not saying it can't be done, only that the air conditioning will not work at normal efficiency.

Heating and cooling systems can be technical and complicated. For example, should people who live in New Hampshire install heat pumps in their homes? Most people would say no, absolutely not. Well, they shouldn't install air-based heat pumps, but water-based heat pumps will work just fine (Fig. 9-8). The cost of installing a water-based heat pump (Figs. 9-9 and 9-10) is high, but the system will work well in very cold climates. If you don't know this type of thing, be careful what you say to your customers. You could open the door to a big problem if you talk about subjects with which you are not experienced.

Should your customers consider earth-source heat pumps (Fig. 9-11)? Can you explain the basic steps in choosing a proper heat pump (Fig. 9-12)? If you are not completely comfortable with advising

Suitability	Good
Stability	Extreme
Availability	Excellent
Initial Cost	Low
Operating Cost	High
Drawbacks	Frosting

9-8 *Profile of an air-source heat pump.*

Suitability	Varies
Stability	Fair
Availability	Limited
Initial Cost	Midrange
Operating Cost	Low
Drawbacks	Corrosion and dry spells

9-9 *Profile of a surface-water-source heat pump.*

Suitability	Excellent
Stabilty	Stable
Availability	Very Good
Initial Cost	Midrange
Operating Cost	Low
Drawbacks	Mineral build-ups

9-10 *Profile of a well-water-source heat pump.*

Suitability	Good
Stability	Stable
Availability	Excellent
Initial Cost	Midrange
Operating Cost	Low to medium
Drawbacks	Leaks are hard to find and expensive to repair

9-11 *Profile of an earth-source heat pump.*

Have a load calculation done for cooling and heating needs

Shop for a desirable brand and type of heat pump

Decide on air terminal requirements

Evaluate outdoor air requirements

Choose duct locations and have ductwork sized

Choose locations for indoor and outdoor equipment

Confirm that installation requirements can be met

Choose equipment controls

Evaluate initial cost of the system

Evaluate efficiency and operating costs for the system

9-12 *Basic steps in choosing a heat pump.*

your customers, have them talk with your HVAC contractor. Giving the wrong advice to a customer is worse than giving none at all.

HVAC systems will not rank high on your list of cosmetic conversions, but they will play a role in some jobs. You should investigate various heating and cooling options before you are asked to explain them to a customer. Get to know what's available and the pros and cons of various systems. Knowledge is most often the key to success.

10

Woodwork

Woodwork in a house says much about the home's character. Whether a home is old and historic with hand-carved trim, or new and fashionable with colonial trim, the woodwork makes a statement. Taking a house that has cheap clamshell trim and fitting it with stained, colonial trim can transform the home into something the residents might not recognize. Decorative trim around windows and doors was common years ago, and it still adds a lot to a house. Chair railing, crown molding, and other types of molding all enchance a room. The finished trim in a house is like the period at the end of a sentence—it completes the statement being made.

There are many parts of a home that contractors don't consider in cosmetic conversions. Woodwork, however, isn't usually one of them. Electrical fixtures might go unnoticed, and HVAC systems frequently are ignored, but most remodelers pay attention to trim work. Is this because trim is part of carpentry work and most remodelers start out as carpenters? I would imagine that that is the case.

If carpenters who have become remodeling contractors notice trim, does that mean it is not noticed by contractors who came from different backgrounds? Does a plumber who opens a full-scale remodeling business pay attention to finish trim? How about the college graduate who enters the field of contracting without any significant field experience? I believe people tend to notice items within their trade specialty more than they do other items. To be a complete remodeler, you must pay attention to all aspects of a job.

Let me ask you another question. I've made my statement of assumption about carpenters being more alert when it comes to trim than remodelers with different backgrounds. I have no proof of this; it's just my opinion. If I'm right, however, is it also true that even carpenters overlook trim options with which they are not accustomed to working? A carpenter who has worked on low-end production housing, for example, probably hasn't installed much chair rail or crown molding. Does this mean that these two types of trim aren't desirable? Of course not. But people do often think in terms of what they are accustomed to and not so much in terms of all the options that are available.

Customers are people, too. If they have grown up in houses with sparse trim, they might not consider adding decorative woodwork to their homes. It may fall on your shoulders to educate consumers in the finer aspects of woodwork. Trying to sell ornate trim to someone who has never lived in a house in which it was installed can be a tough sale. In most cases, the customer will love the woodwork once it's installed, but getting an okay to install it can be very difficult.

How important is woodwork to a house? All houses benefit from trim work. Some homes don't require as much trim as others to remain in step with other houses in the neighborhood. Should only large, expensive homes have ornate trim? Not necessarily, but it is possible to overdo things. Experienced contractors have a good feel for how much is enough and how much is too much.

Regional differences can prove interesting to remodelers who move around. My work in South Carolina was similar to that in Virginia. Not all places recognize the same standards. I recently was talking with a friend who lives and works in Maine. We compared notes on how things were being done in various parts of the country. One topic of our conversation was woodwork.

Shoe molding was one of the issues we discussed. It's common in many places for shoe molding to be installed in front of base trim when sheet vinyl is installed on a floor. This, however, is not the case in Maine. My friend explained that even though he was accustomed to using shoe molding when he moved to Maine, he was in the minority as a contractor. Most contractors in the part of Maine where my friend works don't believe in shoe molding. In fact, many of the contractors don't even know what it is.

The procedure in Maine is to have vinyl flooring installed and then to install base molding over it. This works fine, but it looks strange to a person who is used to seeing the job finished off with shoe molding. Why do some regions use shoe molding and others don't? It appears to be just a difference in the way work is scheduled and done.

It is common practice in many areas for all major trim work, including baseboards, to be installed before flooring is put in. The baseboard is held up to allow carpeting and vinyl to be tucked in under it. With carpeting, no further trim is needed. Tackless strips hold the carpet in place and the bulk of the fiber fills the gap between the flooring and trim. This is not the case with vinyl.

If vinyl is installed after baseboard trim, there is going to be a gap between the vinyl and the bottom of the baseboard. Shoe molding seals this gap and provides additional insurance that the vinyl will not curl or pull loose. Shoe molding also adds a little more dimension to the overall appearance of trim. So, who's right, the people who use shoe molding or the ones who don't? It's not a matter of right and wrong. A case can be made for either method.

Contractors who install vinyl flooring before baseboard is installed save money by not installing shoe molding. They do, however, run additional risk in that the finished floor is subjected to more abuse because all trim work has to be done and painted after the flooring is installed. Shoe molding can be painted or stained before it is installed and tacked into place with minimal risk to new flooring. I favor the use of shoe molding, but the contractors who don't use it have their reasons, and they can't be considered wrong.

Shoe molding is only one example of regional differences. Crown molding is very popular in many areas. According to my friend in Maine, crown molding is fairly rare. So is chair railing, wainscotting, and other expensive trim. Is this because people in one area have different tastes than people in others? Perhaps, but I suspect the reason is money. Adding decorative trim to a house can run the cost up by thousands of dollars, and this cost might not be justified.

I built a standard Cape Cod house for a man several years ago. The man who had me build the house wanted some nice features, such as hardwood floors in the dining room and living room. He also wanted a decorative railing along the stairs that led to the upper level. Instead of going with a simple handrail and square, painted pickets, the homeowner wanted turned balusters and newel posts. He also wanted the treads to be exposed, finished wood. When I first priced the house, the homeowner and I had agreed upon carpeted stairs and a simple railing. He got his upgrade, but his decision cost him an additional $1800. It doesn't take much to run up the cost of a house when you are dealing with fancy trim.

Fingerjoint trim

Fingerjoint trim, that is, trim planed from short pieces of stock joined together, is cheaper than *clear trim*. Once the two types of trim are installed and painted, it's very difficult to tell the difference between the two. Is one better than the other? Clear trim, also called *solid trim*, is stronger than fingerjoint material, but this strength has little merit in real-world usage. The big difference, other than price, is that clear trim can be stained and fingerjoint trim shouldn't be. Notice that I say it shouldn't be, not that it can't be.

I have seen houses, both new and old, in which fingerjoint trim was stained. One can only assume that this was a mistake made by some inexperienced people. I can't imagine that anyone would intentionally want the effect of stained, fingerjoint trim. One new house I saw with fingerjoint trim that had been stained was quite a sight. The window casing and baseboard trim was clear wood that had been stained. Every interior door was surrounded by fingerjoint trim that had been stained. It looked ridiculous. But I can see how it happened.

I imagine the builder ordered prehung interior doors and failed to specify clear trim. The carpenters probably installed the materials they were given to work with, and I guess the painter stained the woodwork that was in place. Slip-ups in ordering and in field supervision resulted in mismatched trim that stood out like a sore thumb. What I can't understand is why the builder didn't remove the trim and replace it with clear trim. Even more interesting is the fact that the home buyer accepted the work.

Fingerjoint trim is so named because it is spliced together with a pattern that looks like two hands joined with the fingers intertwined. As long as the trim is painted, this type of joint is not noticeable. If stain is applied, however, the fingerjoints are very apparent. Take my word for it: You don't want to mix fingerjoint trim with a stain finish. And remember to specify what type of trim you want when ordering trim, prehung doors, or window casing kits. Most suppliers will ship fingerjoint material unless you specify otherwise.

Trim styles

Appearance is a primary function of trim woodwork. There are three basic types of trim used in most houses. *Clamshell trim*, usually the least expensive option, is somewhat rounded and doesn't have any outstanding features. Not too many builders use it these days, but it can still be found in many homes.

Colonial trim probably is used more often that any other type of trim. This is my trim of choice. I've used it in nearly every home I've built or remodeled. No one has ever complained to me about the appearance of colonial trim, and it receives good numbers from real estate appraisers. For my money, there is no better choice for most homes.

The third type could be called *simple boards*. When your customer is after a rustic look, you can use all sorts of trim. One favorite is simple 1 × 3 stock. Some customers like the wood to be smooth, and painted or stained. Others like it rough and stained. The boards can be used to trim doors, windows, and base areas. This type of trim material is inexpensive, in terms of trim material, and provides a rustic look. This, however, is not the type of woodwork that should be used in most houses. A house has to cry out for country charm to use standard boards.

Chair rail

Chair rail, as shown in Fig. 10-1, normally is used only in dining rooms. I have seen it used in breakfast nooks and in some formal living rooms, but dining rooms are the location of choice for chair rail. On a per-foot basis, chair rail is expensive. Fortunately, it doesn't take a lot of chair rail to surround a dining room. If your customer wants a little extra zip in a dining room, chair rail is one way of getting it.

Chair rail can be used with walls that are painted from floor to ceiling. I feel, though, that chair rail is at its best when it separates two different types of wall coverings, such as wainscotting and wallpaper. This is just my opinion, but I would suggest upgrading to at least a combination of wallpaper and paint when installing chair railing.

10-1 *A rustic room with chair rail.* HIP

Crown molding

Crown molding can be used in any room, but formal living and dining rooms are the most likely candidates for it (Fig. 10-2). Crown molding normally is used in specific types of homes. Average homes simply don't employ its use. This is not to say that they couldn't, but they don't. It is important that you consider all aspects of a home before you doll it up with fancy woodwork. If you're not careful, you could create a home that looks very out of place with too much ornate trim.

I've installed crown molding in some master bedroom suites, but most of my installations have been in formal dining and living rooms. It's possible that regional differences might allow for crown molding in other areas, but just don't get carried away with it. Molding should be considered an accent to a home, not a primary feature. If you remember this, you should be okay.

Wainscotting

Wainscotting is not cheap to buy or to install. In fact, it can be downright expensive. Yet wainscotting has many favorable attributes. It gives a formal dining room an air of elegance. Living rooms can come

10-2 *Crown molding in a kitchen.* Wood-Mode

to life with wainscotting, and family rooms can take a lot more abuse without visible damage when wainscotting is installed.

There are many ways to install wainscotting. Some contractors install it with 4 × 8 sheets of material cut in half. Other contractors build the wainscotting one board at a time. The boards might run horizontally, vertically, or in a herringbone pattern. I've even seen old barn boards used to create wainscotting in family rooms.

The primary reason for wainscotting, in most cases, is appearance. It sets off a room and makes a statement. This is the case in formal rooms. Wainscotting has a practical purpose, too. Depending upon the type of material used and the finish applied, wainscotting can protect walls. Let's discuss the elegance angle first, and then we will get into the practical side of wainscotting.

When you are remodeling a formal living or dining room, the wainscotting used is not likely to be old barn boards or fairly inexpensive sheets of wood siding. The material of choice likely will be dress boards, probably with a tongue-and-groove (T&G) fitting. Wainscotting often is installed horizontally for easy nailing, but can be installed vertically, with the use of nailers placed between wall studs. The types of wood used for wainscotting vary. Some extremely expensive versions are used, but pine is a common choice. The wood is often stained, and sometimes painted.

Due to the thickness of some types of wainscotting, a transition must be made where the wainscotting ends and the remainder of the wall picks up. It is common for wainscotting to extend about halfway up a wall. At that point, wallpaper often takes over and chair rail is used to conceal the meeting point of the two types of materials. This makes for an elegant appearance.

When your customer wants wainscotting installed, you are limited only by your imagination. The wood used can be selected in various or even random widths. All sorts of designs can be created. I remember one house, owned by a jeweler, in which the wainscotting had been installed with designs of various jewels made into it. My company didn't do the work, but the workmanship was superb. Personally, I thought the effect was a bit gaudy, but it was unusual and a real eye-catcher.

Many customers never consider putting wainscotting in a rough-and-tumble room, such as a family room where small children play. I see this as an excellent option for avoiding the problems that are so common with painted walls in rooms like these. If you don't have children, you might not be aware of how destructive they can be even when they are on good behavior. Believe me, kids can ruin painted walls very quickly. Wainscotting acts as something of an armor for the low portions of walls in high-contact areas.

I've remodeled many family rooms where the installation of inexpensive wainscotting was a part of the job. The material used normally has been some type of heavy paneling or exterior siding that comes in 4 × 8 sheets. By taking this tough material halfway up the walls, it serves as a backstop for bouncing balls, runaway toy cars, and all sorts of other things that can damage a painted wall on contact. Wainscotting doesn't stop the artistic ability of children with crayons and markers, but it is a lot tougher to dent, scratch, or break through than painted drywall is.

Putting too much wood wallcovering in a room can create a dark environment. This usually isn't a problem with wainscotting. The lower half of a wall might be a little dark, but the upper half can be bright and cheerful. Not everyone will want a wood barrier on their family-room walls, but it never hurts to offer such an option to customers.

Trim kits

Trim kits are available for windows. If you are installing new windows in a home, you might want to consider using these kits. They come complete with all the wood materials needed to trim a window, and the pieces are precut to exact lengths. The kits cost more than it would cost to buy the wood in standard lengths, but a lot of time is saved. You have to determine how much your time is worth in order to decide if the kits are worth their price. In my case, the trim kits have always been beneficial.

Prehung doors

Prehung doors are the only type of doors I use. These doors are shipped with the trim already assembled and ready to use. The split jambs allow easy door installation and fast trim installation. I can't imagine why any contractor would rather fuss with a slab door and then trim it out onsite. Buying prehung doors saves time and the trim work looks just as good as if it were built on the job.

Other wood options

Other wood options exist for builders and remodelers. For example, you could use wood to create interesting designs on tall walls. Intricate wood trim can be added under crown molding to dress a room

10-3 *A dining room with chair rail, crown molding, and a nice piece of woodwork around the fireplace.* Colonial Williamsburg Foundation, Lis King

up. It can be used to dress up a fireplace (Fig. 10-3). Creativity with wood is limited only by the installer.

The finish that is applied to woodwork has a lot to do with how the trim looks (Fig. 10-4). Don't let your painter skimp on finishes when you invest in good wood.

Now that we have covered all of the normal woodwork applications, let's move onto the next chapter and discuss windows, doors, and skylights.

Polyurethane:
>Expensive
>
>Resists water
>
>Durable
>
>Scratches are difficult to hide

Varnishes:
>Less expensive than polyurethane
>
>Less durable than polyurethane
>
>Resists water
>
>Scratches are difficult to hide

Penetrating Sealers
>Provide a low-gloss sheen
>
>Durable
>
>Scratches touch up easily

10-4 *Wood finishes and their qualities.*

11

Windows, doors, and skylights

Windows, doors, and skylights are key components of a home. It has been said that windows are the eyes of a home. Doors provide an essential, practical purpose, but they also can be an attractive part of a home. Skylights brighten rooms, ventilate living space, and decorate roofs. The addition or replacement of these construction materials can make major differences in the cosmetic appeal of a home.

Windows, doors, and skylights can make the difference between an average conversion and a showplace. These items set the pace for an entire habitat. Natural light and good ventilation are critical to pleasant environments. With the proper selection and use of windows, skylights, and doors, the homes of your customers can become fantasy getaways.

This chapter will discuss options available in windows, skylights, and doors. In addition to showing you various product lines, it will give you expert advice on when and where to use the various options. Further, you will be given detailed information on the qualities to look for in your purchases. This information will help you to find your way through the maze of products on the market, to the windows, skylights, and doors that are best for you and your customers.

Windows

Windows are not what they used to be (Fig. 11-1). No, they are no longer just simple pieces of glass held in place by small pieces of wood. Today's windows offer features that can save you money on fuel bills. They are designed to make cleaning easy. You can have windows that push up, roll out, or tilt in. There are even windows that don't require painting, and that's not all. Let's take a tour of the windows you can choose from for your next conversion project.

11-1 *Notice how these windows are made to fit in corners.* Pella

Casement windows

Casement windows are known for their energy-efficient qualities. These fine windows are easy to clean from inside the home, and they offer the advantage of full air flow. When you crank out a casement window, the entire window opens. If you want to smell sea breezes or the fresh morning dew, casement windows bring them to you.

These practical yet attractive windows can be used in almost any application (Fig. 11-2). You can build a picture window setting with them. You can frame a bay window with them and build in a window

11-2 *These windows have blinds installed between the panes of glass.* Pella

seat for your customers to enjoy the view and open air. Decorative transoms can be installed above casement windows to add a touch of elegance (Fig. 11-3). The options for these windows are nearly endless.

Double-hung windows

Double-hung windows, the workhorses of the industry, have endured the test of time. But today the double-hung window has some new features. These features include tilting sashes and removable grids for easy cleaning, along with high efficiency ratings for minimum heat

11-3 *Here a circlehead window has been added over casement windows.* Pella

loss. Add to this vinyl cladding, for low maintenance, and you have a window that is affordable and consistent, in an ever-changing world.

Octagonal windows

Octagonal windows are ideal for bathrooms and stairways. These appealing windows are available with glass that opens or that remains fixed. The glass selections allow for numerous designs and colors. Your imagination can run wild with these handy windows that shed light on the smallest of areas.

Windows, doors, and skylights

Awning windows

Awning windows can set a home apart from the crowd. These windows open out and up, allowing air to circulate, even during gentle rains. When closed, these windows, if grouped together, form a wall of glass. Due to their nature, awning windows can be placed above eye level, making it possible to have fresh air and privacy at the same time.

Bay and bow windows

Bay and *bow windows* are available in premade units. Simple to install, these distinctive windows give any space a new look. With a combination of stationary and movable glass, these window units can be just what your next cosmetic conversion needs. You should investigate all window options and use them to your advantage. Whether you add a bow window or create an enclosure of glass, you are sure to make a house more of a home when you add glass (Fig. 11-4).

Skylights and roof windows

Skylights are ideal for letting light into an attic conversion or a dull kitchen. Whether your customer wants a model that opens or one

11-4 *Glass enclosures can really brighten up a kitchen.* Velux

that remains closed, you will have plenty to choose from when you shop for skylights and roof windows (Fig. 11-5).

What is the difference between a skylight and a *roof window?* In general terms, skylights are mounted in the roof, normally beyond arm's reach (Fig. 11-6). Roof windows, on the other hand, are installed in the side of a roof, to allow views of the grounds and distant scenery. Roof windows are within easy reach and offer many advantages over regular windows and dormers built only to provide light (Fig. 11-7). Let's see how these interesting roof windows might fit into the plans of your next customer.

Roof windows

Roof windows are an excellent choice for attic conversions. They let in a lot of light, yet they can be equipped with window treatments to control the lighting. These windows are available in a multitude of

11-5 *Here are skylights that open for ventilation.* Velux

sizes to suit any need. They are easier to install than a gable dormer, and they allow for broader views and more light.

These versatile windows can be purchased to swing out or tilt in. Some models rotate to facilitate cleaning. Used in a bathroom, the large operable area vents the bathroom and removes moisture quickly. There are models available that are sized to meet the requirements for emergency egress, when used in a bedroom. Most roof windows are available with screens, so your customers can enjoy the open air without insect infestation. All in all, roof windows deserve a serious look for any attic conversion plans or for just brightening up rooms on the upper level of homes.

Skylights

Skylights have matured since the days of the plexiglass bubble that sat atop contemporary homes. These bubbles are still available, and they are a very affordable way to brighten a room. Now, however, the sky is the limit for skylight options.

Modern skylights open, have built-in shades available, and can be equipped with screens. They are available with insulation features not found in older skylights. There are control options available that will close the skylight for you if it starts to rain. With some units, you can use a keypad to control the window treatments and the opening and closing of the glass. Further, there are special rods, some motorized, that allow you to operate out-of-reach skylights. What more could your customers ask for?

Exterior doors

When you begin to look at the options for exterior doors, they may stagger you. Not only can your customers choose from wood, metal, glass, and fiberglass, the range of styles and designs could fill a book. Here, we will look at some of the types and designs that might work best in your cosmetic conversions. Standard entry doors are doors that are normally 3 feet wide and hinge on one side. Aside from this generic description, the options available in entry doors are enormous (Figs. 11-8 and 11-9).

Metal insulated doors

For the price, metal insulated doors are a good choice for most applications. These doors are available as plain solid doors, doors stamped to give the appearance of a six-panel door, and doors made with half the door being glass, with or without grids. Metal insulated doors are affordable and perform well. Most modern builders lean toward this type of door for cost efficiency and energy efficiency.

11-6 *Typical skylight installation.* Summitville, Lis King

Wood doors

Wood doors are capable of providing beauty not found in other types of doors. The carvings and designs on these doors range from the modest to the ornate. Wood doors generally cost more than metal insulated doors, but they do offer looks not possible with a metal door. Further, wood doors can be stained and metal doors cannot.

Wood does have its drawbacks, however. The insulation quality of a wood door will not match that of an insulated door. Wood doors are subject to swelling in damp weather, which can lead to doors sticking or not latching properly. Since your main interest in a cosmetic conversion is creating a dramatic effect, wood doors deserve some serious attention.

Fiberglass doors

Fiberglass doors are available in a host of designs. Some brands can be stained. These doors provide a good wood look-alike appearance, without the swelling problems and with better insulation qualities. Many remodelers do not favor these new additions to the door industry. Like anything new, it takes a little time for old-school contractors to be converted. I suspect that fiberglass doors will enjoy a good market once their virtues become known.

Glass doors

Glass doors are available in various types of frame materials (Fig. 11-10). These doors allow light to flood the interior of a room, but lack insulation and security qualities. Many customers are willing to give up these qualities in order to obtain a fashionable look. While glass doors are not always practical, they are often pretty.

Doors for deck or patio

Doors for a deck or patio can play an important role in any cosmetic conversion (Fig. 11-11). French doors have enjoyed a reputation of prestige for many years. These doors are filled with glass and separated

11-7 *A common roof-window installation keeps the unit close at hand.* Velux

	Exterior	
Thickness	**Width**	**Height**
1¾"	2' - 8" to 3' - 0"	6' - 8" residential
		7' - 0" commercial

	Interior	
1⅜"	2' - 6" min. bedroom	6' - 8"
	2' - 0" min. bath, closet	
Door knob	36" above floor	
Door hinges	11" above floor and 7" down from top of door	
	Optional 3rd hinge ½ way between other 2	
Door clearance	1/16" at top and latch side	
(interior doors)	1/32" at hinge side	
	⅝" at bottom	

11-8 *Typical door dimensions.*

by grills. Both panels of a double French door open. The doors can get very expensive, but they have a distinguished look. They often are installed between living space and a deck or patio, and can be installed to separate a dining room from a living room. The use of French doors is not common, due to their cost, but most customers love the look.

Gliders and sliders

Sliding glass doors have long been known as *sliders*, or patio doors. Sliders are still available, and they are an appropriate choice in some circumstances. As with most products, the quality of these doors makes a big difference in how well they work. Some sliding doors don't

Passageway	Recommended	Minimum
Stairs	40"	36"
Landings	40"	36"
Main hall	48"	36"
Minor hall	36"	30"
Interior door	32"	28"
Exterior door	36"	36"

11-9 *Widths of passageways.*

11-10 *A high-quality glass door with sidelights.* Pella

perform well in cold climates. I've seen sliders on which frost had built up on the inside of the frames in cold weather. Few homeowners want their new doors to freeze up and then melt on their finished floor coverings. During hot weather, houses that combine cheap sliders and air conditioning often experience condensation. Higher-quality doors don't suffer from these types of problems.

Condensation and frost aren't the only drawbacks of cheap doors. Some sliding doors don't slide smoothly. I've seen doors that children couldn't open and that some adults had trouble opening. If you're

11-11 *Hinged, glass patio door.* _{Velux}

doing a conversion, don't stick your customer with a door that will not open easily. Spending twice as much for a similar door might seem stupid during the planning stage of a job, but the end result of buying a cheap door could prove to be a much more stupid move.

In upper-end doors, there are *gliders*. These doors are constructed of high-quality materials and with different techniques from those used in the common slider. Gliders offer more aesthetic options. Gliders can be opened easily with one finger. The cost of these doors can be prohibitive, but their ease of use, appearance, and quality are hard to beat.

Hinged patio doors

Known by many names, the *hinged patio door* is a double-door unit of which only one panel opens. The second panel is fixed and remains sealed at all times. These doors have gained popularity because of their energy efficiency and ease of operation. While looking like French doors at times, these doors are a cost-effective alternative to their expensive counterpart (Fig. 11-12).

Now that you have an idea of the types of windows, skylights, and doors that are available, let's look at some of the technical aspects you should look for when presenting the options to your customers.

11-12 *French-style doors.* HIP

Technical considerations

When you shop for windows, skylights, and doors, there are some technical questions you may want to ask. The following section will get you started on the road to asking the right questions.

Tempered glass If the unit you are buying is to be installed in a location where it could be broken and cut the home's occupants, *tempered glass* should be installed in the unit. Most building codes make this a mandatory requirement. As a responsible contractor, you should share this information with your customers, even if code requirements don't compel you to do so.

R-value The *R-value* is a rating assigned to identify the resistance a material has to heat flow. The higher the R-value, the better insulated the unit is. R-values most often are associated with insulation, but they apply to many types of construction materials, such as siding, drywall, and roofing. Window glass, however, is judged by a different standard. It is known as U-value.

U-value *U-value*, not as well known as R-value, is a rating assigned to determine the total heat flow through a unit. A unit with a low U-value has better insulating qualities than a unit with a higher U-value. When showing your customers various window options, you should stress the U-value of the glass.

UV blockage The *UV-blockage* rating of a unit indicates the amount of reduction in ultraviolet rays passing through the unit. The higher the UV-blockage rating, the better. A knowledge of this type of information could give you an edge over other contractors. The more product knowledge you have, the more likely you are to stand out above your competitors.

Other considerations There are other considerations when comparing units. However, most of these considerations are a matter of taste and money. By looking at cutaway sections and product information, it will be fairly easy to compare products. Each brand will boast its own special features. It will be up to you and your customers to decide which features are worth paying for.

Expenses

The expense of replacing and adding windows, doors, and skylights is sometimes a deterrent to cost-conscious consumers. Cost is something to consider, but there are economical ways to make cosmetic conversions with these features of a home. Something as simple as the installation of a decorative storm door can change the appearance of a house. Adding an octagonal window in a bathroom can make a

Windows, doors, and skylights

tremendous difference in appearance and natural light without busting someone's budget. Even cutting in a cased opening to an existing room can improve appearance and traffic pattern (Fig. 11-13).

Installing new windows throughout a house is a big expense. But how much does it cost to add snap-in grids to existing windows? Not much, and the grids can make the windows look as if they were a new installation.

Repainting a door can be all it takes to create a new look. If a door is beyond the help of paint, replace it. Find a door that will fit the existing jamb and the cost will be minimal. This usually isn't difficult to do. As long as the replacement door is the proper size, it can be of a different style or material. This is an easy way to make a major cosmetic improvement.

Let me give you an example. Assume that you are working with a house where flat, lauan doors have been installed. The doors are old and have been painted white. The wood has started to splinter, and the chrome handles have lost much of their finish. All of the trim in the house is painted white. It's colonial trim that is still in good shape. Only the interior doors are an eyesore. How would you handle this situation?

11-13 *Adding a cased door opening in this kitchen made a major improvement.* Decora

I would go into the home and replace all the lauan doors with six-panel, composition-board doors. These doors, when painted white, will mimic solid wood doors, and they will look great. Installing new hardware along with the doors will complete the picture of perfection. For a builder cost of about $100 a door, or less, you can get the doors and hardware. Installation time is minimal, so the finished cost to a customer should be something that will be considered affordable.

Finding creative solutions to bring boring homes to life can be fun. When you put yourself in the right state of mind, you can enjoy looking for simple, inexpensive ways to alter the appearance of a house. Adding grids to windows, as mentioned earlier, is a great example of how very little work can make a big difference. Windows and doors aren't the only things that creativity can improve. In the next chapter, we will see how you can work magic with cabinets and counters in kitchens and bathrooms.

12

Cabinets and countertops

Cabinets and counters play vital roles in kitchens, and can be prominent fixtures in a bathroom. When new homes are built, these building components typically cost thousands of dollars. They are no cheaper when you are remodeling a house. A straight replacement of counters and cabinets is a major expense. This type of home improvement can be considered cosmetic, but it cannot be considered inexpensive. Yet, the appearance of cabinets and counters has a lot to do with the overall effect of a home's interior. In fact, this phase of work is often at the top of the list for homeowners wanting to freshen up their homes. How, then, can you offer customers the advantage of outstanding cabinets and counters in an affordable manner?

There are several ways to improve on existing cabinets and counters. The most effective way is a straight replacement. If this work proves to be too costly, you can consider replacing doors and drawer fronts. Refinishing existing cabinets is another option, and many homeowners are open to this concept. Refinishing existing cabinets saves not only money, but time as well.

When you tear out the cabinets and counters in a kitchen, you render the room helpless. Simple kitchen duties become nearly impossible. Trying to prepare a meal, store food, or wash dishes can be extremely difficult when working in a kitchen from which the cabinets and counters have been torn out. After all, if these components weren't instrumental in the function of a kitchen, they probably wouldn't exist in the first place.

Kitchens are not the only rooms where cabinets and counters can be important. Many bathrooms depend on vanity cabinets and their tops. Take out a vanity and top, and a homeowner may have to use the bathtub as a lavatory for brushing teeth or washing up. Not many people want to brush their teeth in a bathtub for long. As a contractor, you have pressure from both a cost and convenience point of view when working with counters and cabinets.

Kitchen remodeling is one of the most popular forms of remodeling for homeowners. Statistics show that kitchen remodeling is one of the safest home improvements in which a homeowner can invest. With such a current interest in this type of remodeling, there has been a surge in the cabinet industry. Cabinet manufacturers are offering more designs and styles than ever before. There are do-it-yourself videos available to show homeowners how to reface their existing cabinets. Major retailers are co-venturing with remodelers to offer replacement doors and drawer fronts to be used to give old cabinets face-lifts. When you sit down with a potential customer for a kitchen job, you might have to explain the pros and cons of refacing, replacement doors, and new cabinets. Are you prepared to do this? Let's find out.

Assume that I'm an average homeowner. I've asked you over to discuss the options for improving the look of my kitchen. As we sit at the dining room table, I begin asking you questions. I'm expecting some legitimate answers, so let's see how you do.

I've heard about ways to give my existing cabinets a new look by installing new doors and drawer fronts, so what can you tell me about this procedure? How will I make the rest of my cabinet surfaces match the new doors and drawer fronts? Where will the new hardware come from? Will I have a choice in door pulls and hinges? How much money will I save by having my old cabinets updated instead of replaced?

How are you doing with your answers? The replacement of doors and drawer fronts is a common procedure. This work goes quickly. It can be completed in a day or less, when done by an organized and experienced crew. As for the rest of the cabinet surfaces, they can be covered with a veneer or an adhesive paper that will match the new doors and drawer fronts. Hardware can come from any number of sources. It is available from standard suppliers of building supplies, as well as specialty suppliers. The amount of money saved by refacing instead of replacing can be substantial. An exact amount is difficult to arrive at on a generic basis, but it could easily amount to thousands of dollars.

Now, let's get back to our make-believe estimate interview.

Can you tell me the pros and cons of a wood veneer over a wood-look adhesive paper? How long will a refacing job last? Explain to me the advantages of replacing my old cabinets. How long will it take to get new cabinets? Would you recommend store-bought or custom-made cabinets? Let's review these questions and some potential answers (Figs. 12-1 and 12-2).

Type of Cabinet	Features
Steel	Noisy
	May rust
	Poor resale value
Hardwood	Sturdy
	Durable
	Easy to maintain
	Excellent resale value
Hardboard	Sturdy
	Durable
	Easy to maintain
	Good resale value
Particleboard	Sturdy
	Normally durable
	Easy to maintain
	Fair resale value

12-1 *Kitchen cabinet features.*

If money is a major concern, wood-grain adhesive paper can be used to give a new appearance to old cabinets. This is the cheap way out, and it does come with some potential problems. The paper can be tricky to install. Keeping wrinkles and bubbles out of the paper is not always easy. Since the covering is paper, it can be cut or torn. If this happens, the whole face-lift can be ruined. Sometimes a patch can be made, but more often than not, an entire piece of the paper has to be replaced.

Wood veneer is a better option than adhesive paper when putting a false front on kitchen cabinets. The veneer usually is applied with a contact cement. Since the veneer is made of wood, it looks real. The veneer won't tear or rip, and it passes the touch test. In my opinion, wood veneer is the only way to go when refacing cabinets.

A quality refacing job can last years. It might not be as durable as a new set of cabinets, but for the cost, it is a good value. There are some advantages to doing a complete replacement, instead of a refacing job. When you are forced to work with existing cabinets, your options are limited. Tearing out old cabinets and replacing them with new ones will provide an opportunity for the creative use of all types

Type of Cabinet	Price Range
Steel	Typically inexpensive, but high-priced units exist
Hardwood	Typically moderately priced, but can be expensive
Hardboard	Typically moderately priced
Particleboard	Typically low in price, but can reach into moderate range

12-2 *Kitchen cabinet price ranges.*

of cabinets. Turntable cabinets can be put into a system. Cabinets which house recycling bins can be installed. A number of possibilities exist when you start from scratch.

Timing can be critical when dealing with kitchen cabinets. It can take months to have a set of custom cabinets made and delivered. Stock cabinets, on the other hand, might be available for immediate pickup. Most production cabinets can be ordered and delivered in less than three weeks. All of this comes into play when making a buying decision.

With the many alternatives available in production cabinets, there is little reason to pay extra and wait longer for custom cabinets. The quality of production cabinets has reached a level where it can be difficult to tell a stock cabinet from a custom cabinet. When you weigh out the price and the delivery time of each type of cabinet, production cabinets win the race every time. There are certainly occasions when custom cabinets are in order. If a person wants a unique cabinet arrangement, custom units will be the only way to go.

Counters

Helping your customers make decisions on counters can be almost as difficult as helping them choose cabinets. If nothing else, the multitude of colors and patterns will keep your customers confused for a good while. Will you recommend a square-edge counter or a round-edge unit (Fig. 12-3)? What type of backsplash will you suggest? Are you going to offer to make up a counter onsite, or will you have your customer order a top from one of your suppliers? Is a laminate top the only type you plan to offer your customers? Have you considered a tile counter? How about a tile backsplash (Fig. 12-4)? What about a marble-type counter? Do you think your customer will prefer a slick finish or a pebble finish? Did you know there were this many questions that could come at you so quickly just on the subject of counters? Customers you sit down with might hit you with a lot more questions, and you should be prepared to answer them.

Cabinets and countertops 147

12-3 *A rolled-edge counter.* Westinghouse Micarta, Lis King

Many people don't give much thought to their counters. They go to a showroom, pick out a color and a pattern, and tell their contractor to order it. It's usually a safe bet that a customer will order a butcher-block design in neutral browns and tans. This is the type of laminate top that I'm most often asked for.

If you want to take the path of least resistance, you can do as your customer says and order the top. However, if you want to make a good impression, and possibly a little extra money, you can educate your buyer in the other counters that are available. Many homeowners never stop to think about any type of top other than what they have been used to. If the people have lived in apartments and tract housing, a basic laminate top is what they will assume is standard procedure. Can you imagine how these customers might respond if you show them pictures of nice tile counters? Or, how about a wood top (Fig. 12-5)? If you will be using the job as a reference, it is in your best interest to make the work as nice as possible, and this includes making it a little out of the ordinary. I'm not suggesting that you twist a customer's arm to get him or her to agree to installation of a flashy counter, but it never hurts to apprise people of their options.

Let's assume your customer is dead set on a plain laminate top. Can you sell the idea of installing a tile backsplash between the

12-4 *A kitchen where tile is used as a backsplash.* Rutt, Lis King

12-5 *An elegant counter.* White, Good, & Co.

top and the bottom of the wall cabinets? If you can, there will be a few extra bucks in the job for you, and you will have a better grade of work to use as a reference. The customer will get an easy-to-clean wall of tile, and you will be an above-average remodeler with regard to your base-grade work.

Do you ever build your own counters? If you don't, you should consider it. At the very least, you should find a subcontractor who can do custom, onsite countertops (Fig. 12-6). If you stay in remodeling long enough, you are sure to have an occasion when a site-built top will be of great interest to a customer. I can give you an example of this from one of my recent jobs.

I was putting together a new kitchen package a few months ago. The job involved a corner sink. With the double-bowl corner sink, a stock laminate counter was not a great idea. The seams for the top would meet under the drains of each sink bowl. Not wanting seams where water infiltration might be likely, it was necessary to look for an alternative. In this particular job, a post-form top would have cost me about $300. Since this type of stock top was not desirable, due to the location of seams, I investigated other options. One of these alternatives was a more custom-type top. This counter would be seamed

12-6 *A custom-quality counter.* White, Good, & Co.

away from the sink, but the price for the top would have been a little more than $600. None of these prices included installation.

After weighing my options on prefab counters, I looked into having the top built onsite. Building the counter right in the kitchen would allow me to avoid seams under the sink, and it would speed up the job. Instead of waiting a couple of weeks for an ordered top, I could have one of my subcontractors build the counter in less than a day. The price I was quoted, for labor and material, was less than $450. This option gave me a custom-built counter, which could be installed immediately, for about $150 less than a prefab top that didn't include installation. I chose to have the top built onsite, and it turned out great.

Had I had wanted to, I easily could charged $600 for the counter and a few hundred more for installation. This price would have been in keeping with the cost of a prefab top, and the job still would have been finished quicker. I would have also been about $350 richer. Not being greedy, I didn't inflate the cost of this work. But, I could have pocketed the extra cash without really taking advantage of the customer. The homeowner would have gotten a better counter in less time for the same money as a prefab unit. This example alone proves the value of being prepared to do onsite work with counters.

Bathrooms

When you are remodeling bathrooms, you probably will be installing vanity cabinets and tops. This, of course, is no big deal. Almost anyone can install a vanity base and a molded lavatory top. In many cases, space limitations restrict creativity with bathroom cabinets. But, when there is adequate room with which to work, a vanity can become a work of art.

Most contractors install production cabinets in bathrooms. I've had custom cabinets made for vanity bases from time to time, but a vast majority of my work has been done with stock cabinets. The hardest part about this type of work is helping the customer decide on a particular type or design. When there is enough room to work with, the options for a vanity base are considerable.

If you have only a 30-inch space to work with, you're not going to be able to do much in the way of exotic cabinetry. Such a small space will limit you to a base cabinet with one door and no drawers. There are a few cabinets available in this size with drawers, but they are not common. The larger the space you have to work with, the more you have to offer your customers.

If the space available in a bathroom for a vanity is long enough, you can suggest a double-bowl vanity. Customers can choose base cabinets that have doors under the lavatory bowl and drawers on one side or even on both sides. A fairly standard layout is a cabinet with a door under the lavatory bowl and drawers on both sides. This provides his-and-hers drawer space.

One design that has been very popular with my customers is a base cabinet with an offset top. The cabinet houses a lavatory and the top extends for some distance to create a makeup counter. Add some strip lights, a big mirror, and a seating arrangement, and the homeowner has a fantastic place to get ready for work or play. This type of dressing area is popular in the areas where I've worked.

Vanities can be focal points of bathrooms. The wood, the doors, the shape, and the size of a vanity all can work to create a haven, rather than a simple bathroom. When vanities are used, counters are needed. These can be cultured marble tops with integral lavatory bowls or laminated tops with drop-in or self-rimming lavatory bowls. Self-rimming bowls look better, are easier to clean around, and are not as prone to leaking. Cultured marble tops are usually chosen when a vanity is used. Sometimes, a more expensive type of top is picked, and a few people prefer a laminate top. My experience has shown, however, that a cultured marble top, with an integral lavatory bowl, is a big favorite over other options.

Construction features

Customers may want you to point out construction features as they apply to cabinets and counters. This is not very difficult if you know your product line. What are some of the features that you recommend customers look for in a quality cabinet? Viewpoints on what makes a good cabinet can vary from person to person, but there are some benchmarks with which most contractors agree.

Most contractors agree that good cabinets are enclosed fully, meaning that they have backs in them. Not all cabinets do. Cabinets without backs are not as sturdy as those with backs. How much of the cabinet is made of wood? The more wood that a cabinet contains, the higher it is usually thought of. Were butt joints used in the construction of the cabinet? Dovetail and mortise joints are recognized to be of a higher quality than butt joints. Do the drawers of a cabinet open and close with ease? Drawers should be set into place on smooth glides. If a drawer is jerky to open or close, the quality of the cabinet is probably low. This type of problem could result from a sloping installation, but it is more often the sign of a cheap cabinet. Are the shelves in the cabinet adjustable? They should be. How many adjustment options are there, and how difficult is it to move shelves around? Good cabinets offer a variety of possible shelf heights, and the movement of shelves should be simple.

I believe that customers should see and try cabinets before they buy them (Figs. 12-7 and 12-8). If you are selling stock cabinets, your customers should be able to go to a showroom and compare various styles and types. When custom cabinets are to be made, the customer can't see and touch the exact cabinets that will be used until they have been constructed. But a cabinet builder should have samples of the types of cabinets offered for sale. These samples can provide a glimpse of what to expect from custom cabinets being ordered.

The cost of kitchen cabinets can be extremely high. With so much money on the line, customers owe it to themselves to try before they buy. In other words, they should go out and look at all types of cabinets. Particular attention should be paid to construction features (Fig. 12-9). A cabinet that has terrific eye appeal might be a piece of junk. Getting a cabinet with rotating shelves that don't turn smoothly can be disappointing. Wall cabinets with doors that won't stay shut are a nuisance. Drawers that stick and scrape are no good. You should take your customers to showrooms and let them play with sample cabinets. Encourage them to open and close doors, spin turntables, use drawers, and evaluate all the various types of cabinets.

12-7 *Notice the detail work in these cabinets.* HIP

Cabinets and countertops

12-8 *The beauty of these cabinets is in their simplicity.* Quaker Maid

12-9 *The rich finish and detail work in this material are excellent.* Schrock Handcrafted Cabinetry

While you are at the showroom with your customers, you should discuss door styles and hardware (Figs. 12-10 and 12-11). Will the doors have raised panels? Do leaded-glass doors fit your customer's budget and design? Will the customer want finger grooves or door and drawer pulls? Answer as many questions as you can in the showroom. There is no better place to work through confusion about cabinets and counters than in a showroom.

Balance

It can be difficult to balance a budget and desire at the same time. When I figure a job that will involve new kitchen cabinets, I plug in

12-10 *These cabinets have raised-panel doors.* Congoleum Corp.

12-11 *Flat-panel doors, like these, are sometimes preferred.* _{Mannington}

an allowance for the cost of the cabinets. For example, my proposal will state that my price is based on a cabinet allowance of $2800 (at my wholesale cost). The customer can spend more or less, but since I have no way to know what taste in cabinets the customer has, I have to use a random figure. When I pick an allowance figure, I base it on my gut feeling. Some people spend less than $1500 on cabinets. Others spend $7000, or more. On average, I've found the wholesale cost of cabinets for most of my jobs ranges between $2500 and $3500.

When you take your customers shopping for cabinets, you are likely to see them go through a bit of agony. They know they have a budget to work with, but they have trouble deciding what to do. On one hand, they want to choose something that costs less than their allocated sum. This will save them money on the overall cost of the job. But then, they will have a higher-priced cabinet style that they like better (Fig. 12-12). It may very well be within their budget, but pushing it. Then there will be the cabinets that they really love. These units, of course, are way above the budgeted price. So, what is the customer to do? You may be asked to help settle the dilemma.

Lowball prices

How many times have you seen lowball prices on cabinets in sales flyers? Almost any building supply center that caters to homeowners will carry some type of low-priced cabinets that can be advertised to pull people in. Many times the prices are not the bargain that they seem to be. It is not unusual for the rock-bottom prices of cabinets to exclude finishing. I've had customers come to me with flyers and express a desire for the low-priced cabinets. It is true that unfinished cabinets can sell for half the price of a standard production cabinet that has a permanent finish, but this doesn't make the cheap cabinets a good deal.

12-12 *The glass doors on these cabinets draw favorable attention.*
Quaker Maid

Even if the quality of the cabinets is not lacking, the cost and trouble of finishing them can outweigh any price savings.

I have a policy that stipulates the use of prefinished cabinets. My company will not assume any responsibility for finishing unfinished cabinets. We will install them if a customer insists, but we will not apply the finish. Finishing cabinets to a uniform look is difficult and time-consuming. It is not an area of work which I perceive as being profitable for most remodelers. Unless you have resources beyond those of most contractors, you probably will be better off avoiding unfinished cabinets.

Installation problems

Any time you are working with old houses, you are likely to run into installation problems with cabinets and counters. Floors aren't level and walls aren't plumb. These conditions can make installing cabinets and counters difficult indeed. The problems can be overcome if you catch them early, but when you wait until the cabinets are going in to discover the trouble, you've got a more serious problem.

Anyone who has been a remodeler for long knows that most houses are not perfect. In fact, they are usually far from it. If conditions are not right when cabinets are installed, problems will pop up. Doors might not stay shut. Drawers and slide-out accessories might not work properly. Gaps might exist along the bottoms of base cabinets. The space between a backsplash and a wall can be large enough to drop a pencil through. Face joints along cabinets may not fit tightly. All of this can come together to give any remodeler who hasn't planned for the problems a splitting headache.

When you do an estimate for replacement cabinets, take a good level with you. Check the existing walls and floor. Even if you are going to gut the room, check the existing conditions. Furring out walls and leveling floors take time. If you don't budget this time into your job cost, you're going to lose money. When your level shows that problems with existing construction exist, you should show the trouble to your potential customer. They might not like finding out that additional work is going to be needed to bring existing conditions up to a satisfactory level, but they should appreciate finding out about the added expense before the job is under way.

Setting and hanging cabinets is not difficult work. It's not even very technical. When a room is not square, a floor is not level, or walls are out of plumb, however, the job can become tedious, to say the least. Existing walls can be furred out to make them suitable for cabinets.

A floor that isn't level can be corrected with underlayment and, possibly, some filler compound. None of this work is a big deal if it is done during the rough-framing stage of a job. It can be extremely difficult once the job is nearly finished and it's time to set cabinets.

Very few homeowners are going to appreciate having a wide piece of trim installed along the top of their backsplash to hide a huge gap. A small bead of caulking will be the most that they are expecting. If you have to put trim along the backsplash, someone messed up in his or her planning. It's standard procedure to shim up base cabinets, but the end result should not leave a noticeable gap along the floor. Yet I've seen kitchen after kitchen in which the base cabinets are jacked up to a point where shoe mold is needed to hide the imperfections. This may or may not be acceptable to a homeowner, but having such a gap shouldn't be necessary. If the floor is worked with prior to installing the finished floor covering, there should not be any need for decorative trim to hide gaps.

The key to a smooth cabinet installation is planning. If you prepare a kitchen properly, installing cabinets will be simple and fast. Should you neglect to correct framing problems early on, finishing a job to the satisfaction of a customer can be all but impossible. Spend some extra time up front to avoid conflicts near the end of the job.

Most of this discussion has revolved around kitchen cabinets. This is because they are more numerous than bathroom cabinets. However, the same rules apply for vanities. You need to check closely to see that a new cabinet can be installed properly. You also must allow for the vanity top. This is something a lot of contractors forget to do. They think in terms of a 36-inch cabinet and fail to remember that the top will be 37 inches wide. This extra inch can be enough to cause you to pull some hair out. When you are measuring for cabinets, don't forget to allow for the countertops that will go on them.

Damage

A lot of damage can occur near the end of a job, and some of it is likely to have to do with finished floors and cabinet installations. Using a screw that is too long can result in your having to buy a new cabinet out of your intended profits. If the screw punches through and ruins a cabinet, you are going to be responsible for the damage. Sliding base cabinets into place on a finished floor can cause other damage. If you cut, scrape, or tear a new floor, you will lose more of your profit. Dropping screws and stepping on them is one way that many contractors inadvertently damage vinyl flooring. Some customers may accept a patch, but others will expect a complete floor replacement.

Cabinets and countertops

This gets expensive, so you should make sure that you, and your crews, are careful not to harm new flooring.

Walls often suffer some damage during a cabinet and counter installation. This is to be expected, to some extent. Plan on having your painter do some touch-up work after an installation is complete. Ideally, this work should be done as soon as possible. The visual picture a homeowner gets when a new wall is scuffed or gouged is not conducive to referral business.

To create a complete list of potential problems to be encountered during a cabinet installation would take more space than I have here. For example, a slip in cutting a sink hole into a cabinet can quickly mean a delay of weeks and lost money. A long screw coming up through a new counter is disastrous. Having a worker lean too hard against an open cabinet door can result in damage that is serious enough to warrant the replacement of a cabinet. Almost all of the risks associated with damaging cabinets can be eliminated with good work habits and strong concentration. Take your time and do the job right the first time. You're getting paid for the initial installation, but if you get careless, you might be paying for repairs and replacements.

It seems logical that refacing existing cabinets would be a preference in cosmetic conversions. I have not found this to be true. A majority of my customers have preferred to do straight replacements. Some of these customers probably could have been persuaded to go the refacing route, but I tend to agree with a full replacement if it can be afforded. You may feel differently. Every contractor has to find a comfort level within which to work. I don't have a problem with doing a high-quality refacing job, but I prefer to do replacement work. Find your own niche and excel in it.

13

Accessories

There are many accessories available for interior and exterior applications. Many homeowners are willing and able to install their own accessories, but contractors often do the work. Whether you install accessories or not, you can use them to help convince customers to move ahead with projects. The fun of accessories often prompts people to make changes in their homes. You can use this to your advantage.

What types of accessories are we talking about? They could include anything from a brass doorknocker to an oak towel rack. Glass doors on a shower can be considered accessories, as can rotating tie holders in a closet. Brass kickplates for a front door can be a nice accessory. Stained glass put in or hung in a window can make a difference. As we move though this chapter, I will give you ideas and suggestions for accessories that I've found to be popular. This should help you in your cosmetic contracting ventures.

The exterior

The exterior of a home can benefit from cosmetic and functional accessories as much as can the interior. I can think of one accessory that usually is ignored, but shouldn't be. How many times have you gone out on an estimate and had trouble finding a house? Would it have been easier to locate the home if the house number had been more visible? Of course it would have. House numbers are a very simple, yet very important, accessory for the exterior of a home. It's bad when guests can't locate a home, but it can be deadly if police, fire, and rescue workers can't locate the home. House numbers are available in a variety of styles, sizes, and finishes. This is definitely one accessory that you should consider mentioning to your customers.

Brass kickplates can be to a front door what a necktie is to a business suit. It is the finishing touch that says this house is elegant. Brass door knockers can produce a similar effect when people are close enough to see them. Even the trim around a doorbell button can make a difference in the way a house is perceived.

Mail slots and mailboxes are other items to consider when putting the finishing touches on a cosmetic job. Whether you're doing rural work, and the new mailbox has something of an eggshell finish and a beautiful scene of nature or wildlife on it, or city work, and an old mail slot is being replaced with a new one, this type of extra attention to detail can pay off.

There are other types of exterior accessories that might catch the interest of homeowners. I've had customers ask me to install everything from flag holders to molded American eagles. Accessories of interest vary with the tastes of individual customers. They are the elements of a job that create personality. Contractors can make suggestions to get the ball rolling, but it will be the customers who will set the course for a theme or statement.

Foyers

A home's foyer offers plenty of opportunity for use of accessories. Most of what will be found here doesn't require professional installation. Items might include umbrella stands, coat racks, mirrors, or heavy furniture that provides a place to sit down, hang up garments, and view a mirror. Some type of fancy covers for electrical switches and outlets can be a good idea in a foyer.

Since foyers are typically small and somewhat separated from the rest of the living space, you can stray from the standard materials in the rest of the house. For example, you could dress a foyer up with a fancy light fixture, brass cover plates on the outlets and switches, and some ornate trim, without having to follow this same pattern in the rest of the home.

Living rooms

A living room doesn't often offer much opportunity for contractor-installed accessories. Covers for electrical switches and outlets are a possibility. Attractive poles for drapes and curtains are always something to consider in a living room. Designer blinds for windows are accessories that might benefit a living room.

If a living room serves as a family room, a built-in entertainment center could be in order. The accessories for living rooms vary with the type of living room being outfitted. What works in an informal room might be out of place in a formal setting.

Dining rooms

Dining rooms are similar to living rooms in that they aren't generous with opportunities for accessory options. Some things are the same, such as curtain rods and blinds. Switch and outlet covers are possible accessories, but there isn't a lot to be done with a dining room. Built-in displays for china or similar items might be appropriate. Ornate trim, if you want to consider this an accessory, can be in keeping with a formal dining room. In general, though, there's not a lot you can do in a dining room.

Kitchens

Kitchens are great places for accessories. Living rooms and dining rooms can have you scratching your head for ideas, but kitchens will have you wondering how to decide on what to use. There is just so much that you can do with a kitchen. Accessory items range from under-cabinet lighting to pan racks and space-saver appliances. In addition to visible accessories, there are dozens of possibilities for improving the use of cabinets with slide-out racks, bins, and assorted accessories.

Under-cabinet lighting is, in my opinion, a valuable kitchen accessory. Not only is it functional, it adds to the appearance of a room. This type of lighting doesn't have to be expensive, and doesn't require an electrician for installation. Many types of under-cabinet lighting are designed to plug into a standard outlet. Having the lighting hard-wired is better, but if your budget is small, the plug-in types are better than nothing.

Some contractors don't get into appliances when remodeling a kitchen. Most will install dishwashers and disposers, but a lot of contractors stop at that point. Why? You can make money from the markup on appliances. Don't turn away business. You know people who remodel their kitchens are likely to want new appliances. Ranges and refrigerators are expected in a kitchen. Microwave ovens have become almost a guaranteed appliance in modern kitchens. Here is an opening for you. Suggest a microwave that is mounted under the cabinets. This provides the benefits of the microwave without taking up valuable counter space. The same goes for other types of appliances, such as can openers. Suggest under-cabinet models and supply and install them for your customers. This insures that the job gets done and that when your work is viewed by prospective customers, it will be of professional quality.

Customers enjoy user-friendly kitchens. You can increase the comfort of doing kitchen chores by adding accessories to cabinets. What will you add? The list of possibilities is a long one. Start with a recycling bin. Add a wastebasket pull-out. Install a pull-out cutting board. Vegetable bins are available, as are spice racks and all sorts of customized organizers. Fold-out ironing boards are available. Be aware of what is available and offer the options to your customers. You should make more money. Even if you don't, you will please and impress your customers by giving them the opportunity for a better kitchen.

Bathrooms

Bathrooms are expected to have certain accessories. Toilet-tissue holders are considered mandatory equipment. This simple accessory has many shapes and styles. It can be recessed in a wall or vanity. Oak toilet-tissue holders are popular, as are brass holders. I know some contractors leave the purchase and installation of bath accessories up to homeowners, but I feel this is a mistake. These are the things people see. You should provide and install them to make sure that the rest of your good work gets the acknowledgement it should.

Should you install towel bars or rings? It depends on what the customer wants. In a large bathroom, I would install both. My personal preference is for oak accessories in most bathrooms. Contemporary styles normally look better with chrome or brass. Some customers want toothbrush holders and others don't. The same goes for soap dishes. They can get pretty grungy if they are not cleaned regularly, and some people would rather not have them. Many suppliers offer nice accessory packages that include all the basic bathroom accessories for one low price. I've bought complete oak sets for less than $20.

In addition to the standard accessories in bathrooms, you can add some extras. Coat hooks on the back of a door or on a wall will hold a person's robe. A mirror of some type is pretty much a necessity. Medicine cabinets are not in as much demand as they once were, but many people still like them (Fig. 13-1). Glass doors for showers add to the overall effect of a bathroom, and they aren't that expensive. I think you have to commit to installing the basics and that you should give some thought to helpful extras when reworking a bathroom.

Space can be at a preimum in bathrooms. Consider installing vertical storage units at either end of a vanity (Fig. 13-2). When you are really pressed for space, try using an over-toilet organizer like the one

Accessories   165

13-1 *A lighted medicine cabinet and vertical linen storage unit add to this bathroom.* Quaker Maid

depicted in Fig. 13-3. Bathrooms equipped with pedestal sinks often suffer from a lack of cabinet space. You can change that by installing storage accessories (Fig. 13-4).

13-2 *Vertical storage units.* Quaker Maid

13-3 *Over-toilet organizer.* Quaker Maid

Laundry rooms

Some laundry rooms are so small that they limit the accessories you can install. Others are large enough to accommodate a wealth of accessories. I have found that shelf space is one of the most sought-

13-4 *Storage accessories.* Quaker Maid

after options in laundry areas. Shelves usually can be added to any laundry area, and they don't cost much.

In addition to shelves, you can offer customers fold-down ironing boards and tables. The tables come in very handy when sorting clothes. Your plumber probably could install a laundry tub without much trouble or expense. The tub can be used for soaking clothes, wringing out mops, or washing family pets, among other things. Built-in hampers are another option to consider for laundry areas, as are drying racks. Something as simple as a paper-towel holder can come in handy in a laundry room. Since many people use their laundry areas as storage facilities, wall-mounted racks for brooms and mops could be useful. Depending on the size and location of a laundry room, there are many potential accessories that can be used.

Studies

A study or den frequently can be improved by adding bookshelves. If the room is used as an office, the installation of built-in desks and computer stations can be helpful. File cabinets can be installed in the base cabinets used to hold bookshelves or office equipment. Trophy cases sometimes are sought for studies, and these can be built-in units. Magazine racks are another option that can be useful in a study.

Some rooms, like studies, have fireplaces in them. If the fireplace is old and scarred from fires, you can dress it up with tile (Figs. 13-5 and 13-6). Yes, tile can be used to enhance the apperance of old fireplaces. This is not a common cosmetic conversion, but it is one that will attract attention. There are many types of tile designs available (Fig. 13-7).

Bedrooms

Bedrooms are often left very plain. A typical bedroom has four walls, a ceiling, and a floor, without much else. This doesn't have to be the case. Consider building an entertainment center into the room. Maybe a custom-made stand for a large fish aquarium would make your next customer happy. Bookcases and window seats are both popular bedroom additions. One area that you might want to concentrate on is the closet. Would your customer enjoy a closet that has mirrored doors? Many people do. Can you build in shoe racks or shelves to improve the organization of a closet? You probably can, and all homeowners appreciate having more usable closet space. A faux fireplace can add a little romance to a bedroom. While it is typical to keep bedrooms plain, there is no rule that says you must. Get

13-5 *A fireplace that has been dressed up with full tile coverage.* Country Floors, Lis King

creative. Design some ideas that will make your work stand out, and you should get more jobs.

Accessories can make any room more attractive and more comfortable. Manufacturers make it possible to buy stock accessories for almost any need or desire. Get on the mailing lists of some specialty manufacturers. You can find many of these sources in monthly magazines devoted to home improvements. Order the catalogs that are available to you. Until you become aware of the myriad opportunities available, you can't offer them to your customers, and this could be costing you a bundle of money.

13-6 *This fireplace has been accented with tile.* Summitville Tiles, Lis King

Chapter Thirteen

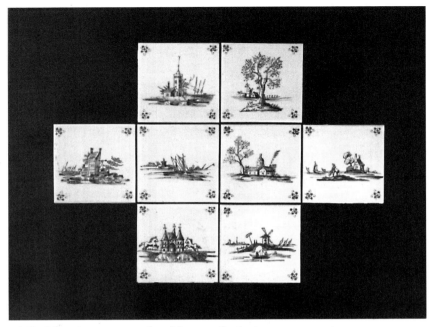

13-7 *Here is an example of fancy tile designs your customers can consider.* Summitville Tiles, Lis King

14

Kitchens

Kitchens will be the topic of conversation in this chapter. Until now, cosmetic conversions have been discussed in an overall sort of way. It's now time to get specific. This chapter, and each of the following chapters, will concentrate on particular rooms. You will be given lots of ideas for effective cosmetic changes. By doing this on a room-by-room basis, you should find plenty of ideas for any type of room that you may be asked to remodel.

The goal in this book is to discover ways in which an ordinary room can be turned into something extraordinary. You will learn to do this with inexpensive items, and with work that is not so inexpensive. Some homeowners will consider painting their kitchen and replacing the kitchen sink a major expense. Other property owners will think nothing of gutting a kitchen and rebuilding it from the studs and subflooring. You need to be prepared for both types of customers. Let's have some fun and see what creative ideas we can come up with for the transformation of kitchens.

Floors

Kitchen floors take a lot of abuse. Most kitchens are high-traffic areas in which spills are likely, a frying pan occasionally hits a floor, and cans of food and other items are dropped. It's not uncommon for kitchen areas to be the first part of a home that residents enter. Think about it. How many kitchens can you remember seeing that led directly into a garage? Builders often position kitchens close to garages and parking areas to make it easy to unload and put away groceries. Foot traffic coming in from the outside can bring a lot of dirt into a kitchen area. Homeowners with children often allow them to play in the kitchen area while chores are being done. A child racing toy cars across a floor can leave scuff marks and scratches on the flooring. All in all, kitchen floors often live a hard life (Figs. 14-1 and 14-2).

14-1 *Here is an example of a tough kitchen floor that will stand up to abuse.* Wood-Mode

Some types of floors are more durable than others. Since most modern kitchens have vinyl flooring, the floors are susceptible to cuts and burns. The sharp edge of a can or frying pan can easily slice a vinyl floor. Vinyl flooring that has been in place for years often becomes dull. Grit on a vinyl floor can result in depressions in the vinyl as it is walked on. And old flooring frequently shows its age just by its pattern or color. All are good reasons to consider upgrading the floor of a kitchen during a cosmetic makeover. The various types of flooring that can be used in kitchens were discussed earlier. This is a

14-2 *This floor is not only elegant with its designs, but it's tough, too.* Armstrong World Industries

good time to consider the practical use of those materials. Let's see when it might be best to use specific types of materials.

Vinyl flooring normally is the most sensible type of flooring to install in a kitchen. Light colors make a room appear larger and more inviting. Picking a flooring with a light, but bright, color will give the impression of cleanliness. This is an important message to send when dealing with a kitchen.

Patterns in a vinyl flooring can change the mood of a kitchen. Homeowners should give careful consideration to the patterns they choose. A very busy pattern might make dirt harder to see, but also can make a room appear cluttered. Since some kitchens are small, a pattern that offers an expansive feeling is frequently a better choice. Mixtures of colors can work to a kitchen's advantage, or against it. For example, a routine mix of white and black squares might look fine in a large kitchen, but out of place in a narrow, galley-style kitchen. Flooring styles with basic bright colors and containing flakes of color are universally acceptable, but can be boring. I prefer some type of pattern in the flooring my company installs.

Wood flooring (Figs. 14-3 and 14-4) is expensive to buy and install. I don't like wood flooring in a kitchen, with some exceptions for country kitchens. I've had customers request wood flooring in kitchens; on the whole, however, sheet vinyl is the type of flooring most often used. Except for special circumstances, I believe a remodeler and a customer are likely to do best with vinyl.

Walls

The walls in most kitchens are painted drywall. White and off-white are common colors. Years of cooking can give these wall surfaces a brownish tint. The change is subtle, and homeowners rarely notice the change in their wall colors. Visitors, however, frequently can see the discoloration immediately.

Walls typically are one of the first components of a kitchen to get attention. It's not unusual for homeowners to paint the walls themselves. But they generally don't prepare the wall surfaces properly, and their paint jobs might not last. The moisture, grease, and smoke associated with kitchens are hard on paint. All of this leads to good reasons for refinishing existing kitchen walls.

I like to use tile on kitchen walls between the counter and cabinets. Tile is such a versatile material. It's easy to clean, resists grease and oil, and provides a beautiful backdrop for a kitchen. Design options are nearly unlimited. A customer can have a tile wall of vegetables, country scenes, or brightly colored patterns. Any decor can be complemented with ceramic tile. This is an excellent way to make a kitchen look completely different in before-and-after photos.

Wallpaper is another good option for kitchens (Fig. 14-5). The multitude of colors and designs make it possible for a designer, remodeler, or homeowner to create thoughtful themes. People often shy away from tile and wallpaper because of the cost. When you think about it, though, most kitchens don't have a lot of exposed wall space. Most of the wall area is covered with cabinets, so this makes a kitchen the perfect place to invest in a higher grade of wall covering than paint.

I've seen some kitchens redone so that their walls were textured. One kitchen that I recall was stippled, but most of the textured walls I've seen in kitchens have been struck with trowels. This can create a pleasing effect, but I would think that textured walls in a kitchen would act as magnets for dirt and grease. Since I've never textured a kitchen wall for a customer, I can't comment on how well the finish holds up or how happy customers are with it.

14-3 *Wood floors can make a kitchen dark; offset this with light-colored cabinets.* Rutt, Lis King

14-4 *Wood squares are a popular option for flooring.* Wood-Mode

Stenciling is an inexpensive way to put a finishing touch on the walls of a kitchen. This can be done with a border-type wallpaper or with actual stenciling. Both achieve a similar effect. I've seen stenciling done with subjects ranging from pineapples to songbirds. Done tastefully, stenciling is a prime consideration in a cosmetic conversion.

Ceilings

Kitchen ceilings tend to be flat and normally are made with drywall covered with paint. There's nothing wrong with this, but some

14-5 *Wallpaper accents this unusual kitchen dining and pass-through arrangement.* Pella

kitchens cry out for more. For example, the installation of false exposed beams can bring a kitchen to life. Beams don't work well in all kitchens, but there are times when they are the perfect decoration. I don't feel that texturing is a good idea for a kitchen ceiling. The texture grabs every piece of dust, dirt, and grease. It's nearly impossible to wash a textured ceiling, so I can't see using one in kitchens.

Adding skylights in the ceiling of a kitchen brings a wonderful addition of natural light and enjoyable views. Being able to look up from kitchen chores to see blue sky and white puffy clouds can take some of the drudgery out of kitchen work. If the skylight can be opened, it adds ventilation to a room that often can benefit from a good draft.

Windows

Windows can be vital to the appearance of a kitchen. Many kitchens have only one small window, usually located over the kitchen sink (Fig. 14-6). This is adequate, but not enthralling. Replacing this type of window with a nail-on garden window changes the entire personality of a kitchen. Garden windows are not inexpensive, but for

14-6 *A wall of windows will brighten any kitchen.* Lis King

customers who can afford them, they work wonders. Adding a garden window not only provides a strong visual point, but it allows more light to flood a kitchen. And this can make a room feel much larger.

Wall space is one reason why kitchens generally are limited in their window placement. This can be a difficult problem to work around. If you will be replacing existing wall cabinets, you might be able to design a new layout that will accommodate a larger or extra window. Getting light into a kitchen is important, and the added ventilation never hurts.

There is another possibility for getting natural light into a kitchen when most of the outside wall is covered with wall cabinets. The space above wall cabinets can be used to house either fixed-glass panels or small windows. Cutting these units in is not very difficult, and the result can be surprising in a nice way. Try to envision a kitchen in which the upper wall section is filled with glass panels fitted between the wall studs. You can use clear glass or frosted glass blocks, depending upon the effect you are after. This type of improvement doesn't cost a lot, but it gives a typical kitchen a chance to be much more interesting.

Doors

Many kitchens don't have doors to separate them from other living space. When kitchens have exterior doors, they often are plain and simple. If a kitchen has an exterior door, it can be replaced with a door that has top-to-bottom glass, or at least an upper half filled with glass. In some parts of the country this could create enough of a security problem that it wouldn't be feasible. But, when possible, installing a glass door will brighten the room tremendously.

Interior doors for kitchens often are nonexistent. The opening between a kitchen and living space is often left unencumbered by a door. Why is this? There probably are many reasons, but I have a few ideas on the subject. Space is one consideration. Since many kitchens are cramped, creating room for a door to swing would only add to the problem. Doors on a kitchen could create an atmosphere that would make residents feel closed in. I suspect that doors are omitted to give a more open, airy feel to kitchens. All of this is well and good, but what about people who don't want guests staring into their kitchens from other parts of the home?

Have you ever noticed that a lot of houses are designed with their living rooms adjacent to the kitchens? This is a common plan. Does it make sense to have an open kitchen in easy view of guests? Even if you are a fastidious housekeeper, a kitchen usually isn't the most desirable background for entertaining. For people who let dishes pile up in a sink or on a counter, an open kitchen can be an embarrassment. What's the answer?

I've often been asked to screen a kitchen from other living space. Due to space restrictions and customer desire, the use of traditional doors normally has not been an option. I've found two effective ways to solve the problem. One involves more work and money than the other, but both ways work. My preferred method is to install a pocket door. I have one of these in my house, between the kitchen and the living room. The advantage to a pocket door is that it takes up almost no usable space, can be left open under normal circumstances, and closed when situation warrants. This is a perfect solution, except for the work that is involved to install such a door.

My second option is the use of swinging doors, like the ones shown in saloons in old television westerns. These doors require very little space in which to operate. They allow plenty of light to circulate above and below them. And they can be hung at a height that allows them to screen the counter area of a kitchen, which is usually the most cluttered part. Swinging doors are affordable to purchase and simple to install. They provide an almost instant screen with no destruction to existing walls or ceilings. For a quick fix, swinging doors are hard to beat.

Cabinets

Cabinets are a big part of most kitchens. They also are an expensive part. As cabinets age, they often deteriorate. Drawers don't open like they should. Cabinet doors stop shutting well. Scratches develop on the cabinet finish. Styles change and cabinets become outdated. This, among other things, leads homeowners to make a change. Maybe the only reason for a change is to get away from dark cabinets and replace them with lighter colors. Some customers want substantial upgrades, with leaded-glass doors. In older homes, customers want the convenience of pull-out shelves and rotating storage. For houses that are very old, new cabinets are needed to create more storage space. There are lots of reasons for refinishing and replacing cabinets.

Many good cases can be made for refinishing instead of replacing. If the only reason for a change is cosmetic, refinishing makes a lot of sense. In all my years as a remodeler, however, I've rarely been asked to refinish a set of cabinets. Most of my cabinet work has involved replacements and additions. This could be because I don't specialize in refinishing and I don't advertise for this type of work. There are companies that appear to do nothing but refinishing and refacing, so there must be demand for the service.

In terms of cost, a customer can't get a new look from cabinets for any less money than will be spent on a cosmetic refinishing or refacing. The financial savings over a full-blown replacement can amount to thousands of dollars. As long as the makeover is done with quality materials and workmanship, there is nothing wrong with saving money.

My parents have been considering the replacement of their kitchen cabinets. Their house is fitted with very dark cabinets. They have lived with them for years. This is partially because they have chosen to attack other elements of the home in their improvement efforts. For example, they replaced all the ugly, green carpeting and drapes with bright, delightful colors. They also replaced the vinyl flooring in their kitchen and made many other cosmetic improvements.

When my father asked my opinion on the cabinets, I inspected the units. Structurally, the cabinets were sound, with the exception of the doors. I suggested that he have the cabinets refaced with a wood veneer and replace all the drawer fronts, doors, and hardware. If I'm willing to give this advice to my parents, I certainly wouldn't hesitate to recommend the option of refinishing to regular customers. But there are times when refinishing just doesn't make sense.

I inspected the cabinets in a house a little over a year ago for the possibility of refinishing. The exterior of the cabinets was a mess. Unfortunately, so was the interior construction. Apparently, the cabinets were a cheap grade that had been treated with very little respect. After assessing the condition of their construction, I advised the owner to do a complete replacement.

In most cases, customers who call me for kitchen remodeling want new cabinets. If I'm talking to customers who tell me what they want, I'm not going to try talking them out of their decision, unless I can see that they are making a serious mistake. Maybe this is why my company doesn't get much refinishing work. If you like the idea of doing quick in-and-out remodeling with refacing and refinishing, I suggest you advertise specifically for this type of work. If my personal experience proves anything, you won't get a lot of demand for refinishing as a general remodeler.

Counters

Counters are one element of a kitchen that probably takes even more abuse than floors. This often contributes to the need for early replacement of countertops. Counters and cabinets normally are installed at the same time; if one needs attention, the other probably does too. There have been a few occasions, however, when my remodeling work was limited to just counters and plumbing fixtures.

Counters are another area in which contractors can make a lot of difference in the appearance of a kitchen for less than $1000. Straight replacements are the most common type of counter cosmetics, but you can do more or do less. For example, you might strip the laminate from an existing counter and recover it. You could install a new top that protrudes far enough to create a breakfast bar, assuming there is enough floor space available. Other eating arrangements can be made with the creative use of counters (Figs. 14-7 and 14-8).

Do you remember the old kitchen counters that were surrounded with chrome strips? Would you believe some of these antiques are still in use? Well, they are. When you go on an estimating call for a kitchen overhaul, you never know what you will find. I've found slate counters with deep, heavy sinks. Then I've run across old wooden counters and original wood cookstoves. Some of the kitchens I've worked in were still equipped with old pitcher pumps that supplied water to a deep sink.

14-7 *The simple addition of a counter has turned this kitchen into an eat-in kitchen.* Quaker Maid

Most kitchen counters in modern homes are made with laminate tops. These tops are efficient and relatively inexpensive. The last stock countertop I priced ran about $300, at my cost. This isn't a lot of money when you consider how much difference a new top can make in the look of a kitchen. Some laminate tops cost much more.

I've used tile to make kitchen counters on many jobs. The first home I built for myself had a tile counter. I like tile because of the many decorative options available with it. Tile is extremely durable under average conditions. There is some risk that the grouting will age and begin to leak, but this is about the only drawback I can think of for a tile counter.

In addition to laminates and tile, there are some other extremely nice counter materials available. Cost can become a problem with these special types of tops, but the looks are hard to beat. One type of top that I've used is so tough that a fine-grit sandpaper can be used on it to remove deep stains and burn marks. Regardless of which type of counter your customer chooses, the replacement of old counters is definitely a step in the right direction towards reaching a cosmetic goal.

14-8 *Good design and the use of a counter has created a cozy cafe atmosphere in this kitchen.* Feincraft, Inc., Lis King

Plumbing

The plumbing in a kitchen can present some appearance problems. Old sinks and faucets often show their age. Replacing a counter without replacing these fixtures should be a crime. It's not, of course, and some contractors do reinstall existing fixtures in new counters. When you consider that a new stainless steel sink and faucet can be bought at wholesale prices for less than $150, I can't imagine why anyone would reuse old fixtures.

I will admit, however, that sometimes an existing sink is worth keeping. This can be the case with any type of kitchen sink, but it is especially true of cast-iron sinks, which are expensive and which usually hold up well under heavy use. When you look at a job, make sure you assess the plumbing fixtures to determine if they should be replaced.

Electrical

Electrical improvements in a kitchen are limited. If your customer is looking only for cosmetic improvements and doesn't want to go to a lot of trouble, you will be restricted to working with existing fixtures. If the customer gives you a little more latitude, your electrician can do much more. For example, you could install lighting under the wall cabinets. This creates a wonderful effect visually and it makes working in a kitchen easier and safer. Track lighting can be added to brighten up a dark kitchen, and recessed lighting can be installed to provide accent lighting (Fig. 14-9). Depending on the style of the kitchen, a ceiling fan could be a handsome addition.

If your electrician is allowed enough freedom, make sure the outlets in a kitchen are up to current code requirements in spacing, amperage, and ground fault circuit interruptor (GFCI) protection. Taking the time to upgrade concealed wiring might not add to the cosmetics of a kitchen, but it does add to its safety and convenience.

Appliances

Appliances might be something you don't get involved with. But appliances can have a lot of impact on a kitchen's looks. Something as simple as changing the cover panels on a dishwasher can make quite a difference. Many dishwashers come with assorted panel colors from which to choose. If your customer is going from black appliances to something in an almond color, you might be able to switch the

14-9 *Recessed lighting helps to brighten this kitchen.* Millbrook Custom Kitchens, Lis King

cover panel on a dishwasher and make it match the new range and refrigerator.

I don't normally consider appliances a part of my responsibility as a remodeler. As a builder, I often provide appliances with new homes, but not on remodeling jobs. However, I do counsel my customers on the benefits of replacing old appliances with new ones. Harvest gold refrigerators and avocado green ranges are a thing of the past. Get rid of them. Rundown appliances can take the shine off a new kitchen. Help your customers help themselves by educating them in good remodeling practices.

15

Bathrooms

Bathrooms probably are second only to kitchens as the room most frequently targeted for remodeling. This includes decorative and cosmetic work, as well as full-blown remodeling. Surveys show that bathroom improvements are one of the best investments a homeowner can make when spending money to fix up a house. There is plenty of supporting documentation available for remodelers to prove to homeowners that bathroom improvements can be advantageous.

Average bathrooms are small in comparison to other rooms in a house. But within these small spaces are some expensive components. It's not just the cost of what occupies space in a bathroom, but the cost of plumbers, who usually are required when work involves major bathroom remodeling. If you computed cost on a per-square-foot basis, I think you would find bathrooms to be one of the most expensive rooms in a home. You've got a little room with a big price tag, so where does this leave you?

Bathroom remodeling can get expensive. I've done remodeling work in bathrooms that cashed out at more than $8000 just for the plumbing. It's common for homeowners to spend several thousand dollars to have a bathroom updated. When a complete remodel is in order, the cost soars, usually in the range from $5000 to $12,000, at least in the areas where I work. For most people that's a lot of money. Is it possible to make improvements that won't force homeowners to take out a second mortgage? Sure it is. There are lots of ways to dress up a bathroom without spending a huge sum of cash. Let's look at some of them.

Floors

Let's start with the options available for floors in bathrooms. Most bathrooms have some type of vinyl flooring (Fig. 15-1), but a good many have tile floors (Fig. 15-2). Both of these types of floor are used in modern construction, and either will provide years of dependable

15-1 *Vinyl flooring that gives the appearance of tile.* Mannington

service. Vinyl flooring is easier to replace than tile. For obvious reasons, removing tile is more difficult than removing vinyl. In some cases, vinyl doesn't have to be removed before a new layer is installed.

Flooring is not cheap, but bathrooms tend to be small. You and your customers have several things to consider when it comes to doing something different with a bathroom floor. Assume that a bathroom has an existing tile floor. The grout between the tiles has discolored and deteriorated. Maybe all you have to do is replace the old grout to create a new look. The new grout coud be a different color, giving the existing tiles a different appearance.

Depending upon the type of tile installed on a bathroom floor, there might be no reason to replace it. Tile typically wears well. Unless there is something about the tile that screams out the age of a house, you probably can leave the flooring alone. An exception to this is, of course, the times when the homeowner wants a new floor for a new look.

Bathrooms

Let's say you are working with an existing vinyl floor. The old floor is covered with vinyl squares, some of which have started to curl. This is a case in which the flooring should be removed and new flooring installed. But suppose the floor is covered with a respectable sheet vinyl that is in good shape. Is there anything that you can do to boost the appeal of the flooring? Suppose you remove the old baseboard trim and replace it with a tile border? This approach would create something a little unusual and probably would capture the attention of visitors. Something as simple as a tile border, instead of baseboard trim, could make a big difference.

15-2 *A real tile floor.* Tubmaster

Few houses being built today have tile floors in the bathrooms (Fig. 15-3). Cost is a major reason for this, and safety is another. Tile often gets very slippery when wet. Comfort is another factor. Tile floors tend to be cold to the touch. Since a lot of people don't wear shoes or slippers in a bathroom, a cold floor can be a real drawback. For most jobs, I would stick with sheet vinyl as the flooring of choice.

15-3 *Designer tile work gives a look of elegance to this bathroom.* Armstrong World Industries

Walls

The walls in bathrooms can be painted, tiled, wallpapered (Fig. 15-4), covered with water-resistant paneling, or decorated with some other type of covering. In many older bathrooms, tile runs about halfway up the walls. A lot of customers want the tile torn out and replaced with wallpaper or paint, but others want tile extended to create a full tile

15-4 *Wallpaper is often a good choice for a bathroom.* HIP

wall (Fig. 15-5). You can't be sure what people will want from one job to the next.

Bathroom walls tend to be a target for cosmetic improvements. Tiled tub enclosures are replaced with plastic or fiberglass enclosures. Old pink and black wall tiles are torn off walls and replaced with paint or wallpaper. Glass blocks are used to make partitions (Fig. 15-6). Murals are used in some bathrooms. Unfortunately, cheap water-resistant paneling is sometimes installed over painted walls. There are numerous ways to change the walls of a bathroom. Since there are so many options, let's discuss them individually.

You can't go wrong with painted walls. While painting is a simple wall finish, it is well accepted in nearly every real estate market. A fresh coat of paint can make a bathroom look much better, and it is a cheap alternative to other, more costly methods of wall covering. Paint can be boring, however. People are used to seeing painted bathrooms, so they are not likely to stop and stare in awe at fresh paint. That's the downside.

15-5 *Full tile walls.* Tubmaster

15-6 *Decorative glass blocks in use as a wall.* Mannington

Wallpaper can get expensive. Even so, wallpaper is an excellent choice for customers who don't want a run-of-the-mill bathroom. Decorative options in bathroom wallpaper are extensive (Fig. 15-7). Some people run wallpaper halfway up a wall and then paint the upper portion of the wall. Molding is used to separate the two types of coverings, creating a somewhat formal look that appeals to many people. A few people will argue that, because of the moisture content of the room, a bathroom is not a fit place for wallpaper. As long as the wallpaper is rated for bathroom use, there shouldn't be any problem with such an installation. Personally, I like the idea of using wallpaper in nice bathrooms.

The demand for tile walls isn't what it once was. I think the expense is part of the reason tile has declined in popularity. I also think that as children from different generations grow up and become homeowners, they don't consider tile to be the mandatory bathroom wall covering it

15-7 *Wallpaper can be used to achieve outstanding results in a bathroom.* Congoleum Corp.

once was perceived to be. I can't remember the last time a customer asked me to install tile as a wall covering in a bathroom, except for bath and shower enclosures. I tear a lot of tile out, but I don't put much in. Maybe my personal experience isn't reflective of the market on a whole, but from where I stand, people are losing interest in wall tile, at least in bathrooms. I do continue to get a fair demand for wall tile in kitchens.

Most of the tile my company installs in bathrooms is used as an enclosure for a tub or shower. Even the demand for this work has dropped off considerably. More and more people are having their tile enclosures removed and replaced with plastic or fiberglass. I much prefer the fiberglass enclosures, but we will talk more about them in a moment.

Some customers have asked me to install tile along the top of their bathroom walls, as a border between walls and the ceiling. The

Bathrooms

effect is similar to stenciling, and it looks nice. I've also used tile as a divider between paint and wallpaper when a wall is done half and half. A border of decorative tile makes for a nice transition from wallpaper to paint.

I've touched on shower enclosures, but let's dig a little deeper on this subject (Fig. 15-8). I can't count the number of bathroom remodeling jobs that have started for me with a leak in a downstairs ceiling. Water running through deteriorated grout, between the wall tiles surrounding a tub or shower, can do a lot of damage. Once the cause of

15-8 *A corner-shower enclosure.* Tubmaster

this type of leak is identified, few homeowners opt to have the tile regrouted. They usually want it replaced with new tile or some other type of surround.

There are many types of plastic tub surrounds on the market. Fiberglass surrounds also are available. Many of these materials are approved for installation over existing tile walls. In all my years as a remodeler, I've never installed a surround over existing tile. I insist on removing all tile, and often the drywall behind it, before installing a new surround. It's important to me that the new surround be installed on a clean, dry surface. Given the choice, I replace old drywall with water-resistant drywall and then install the surround. A few types of surrounds are meant to be installed tight against stud walls. With these models, you have no choice but to remove all existing wall materials before an installation.

I have used one brand of fiberglass surround for more than 15 years. In all those years and countless installations, I have never had a problem with the walls leaking or falling off. I refuse to install cheap plastic surrounds. If I'm going to stand behind my work, and I do, I insist on using quality materials that have a proven track record. The downside to my position is that the walls I use wholesale for around $285, nearly as much as a brand new, one-piece tub-shower combination. If you look in the flyers distributed to homeowners from various home-supply stores, you can find complete tub-surround kits being sold for less than $40.

I honestly can't say how good or bad the cheap plastic walls are, since I've never installed any of them. It stands to reason that they should be waterproof, but I don't know how well they hold up after installation. As a professional, I stick with professional products that I know I can depend on.

A tub surround is an affordable way to make significant changes in a bathroom. For example, you can take a bathtub that doesn't have shower facilities and turn it into a tub-shower unit by installing a tub-shower valve and surround. This can mean a lot to a homeowner, and it doesn't have to cost much, especially if you are willing to gamble on inexpensive, plastic tub surrounds. If you're not familiar with various types of surrounds and the features they offer, you should spend some time getting educated on the subject. Talk to your suppliers and your plumbing contractor. Tub surrounds are a fast, easy, cost-effective way to make a sizable difference in a bathroom.

I mentioned earlier that some people put murals in their bathrooms. This doesn't seem to be as popular as it was a few years ago, but I still get calls for this type of work. Depending on the layout of a bathroom, a mural can make a tremendous difference in the room.

You should consider murals as options to other types of wallcoverings, but don't get too carried away with them. Most murals work best when limited to just one wall. Doing all of the walls in a room with murals would probably be too much.

Ceilings

Bathroom ceilings don't offer a lot of opportunity for cosmetic makeover. Painting or texturing is normally about the extent of feasible options. I have installed false beams in some bathrooms, and I've installed tongue-and-groove planking as a ceiling, but this type of work usually isn't justified. I must say, however, the tongue-and-groove ceilings I've done have always looked great. They have always been done in spacious bathrooms, so I'm not sure how they would work out in a small room.

If you've been a remodeler for long, you know that bathroom ceilings sometimes become traps for mold and mildew. This is the result of inadequate ventilation. If you run into a ceiling where mold and mildew are problems, don't think that a sealer and fresh coat of paint will solve the problem. You have to solve the ventilation problem before the other problem will go away.

Plumbing fixtures

Plumbing fixtures are important to the appearance of a bathroom. Old fixtures don't give a bathroom a modern look, no matter how much other remodeling you do. If the fixtures in a bathroom are old, stained, or cracked, your customer should have them replaced or refinished. New plumbing fixtures can be expensive, but many fixtures are available at reasonable prices.

If you are working with an older home, you probably will be dealing with a wall-hung lavatory supported by two chrome legs. Don't be surprised if it's pink. This type of lavatory once was standard procedure in a lot of bathrooms. Replacing such a fossil with either a pedestal lavatory or a vanity and top should be considered essential. You usually can have either a pedestal lavatory or a vanity and top installed in the same amount of space allocated for a wall-hung lavatory. Vanities provide storage that is not available with a wall-hung lavatory, and pedestal lavatories add a touch of class to a bathroom. Either option is better than a wall-hung model.

You might find old toilets, or at least their tanks, hung on the wall of a bathroom. It's rare to find a completely wall-hung toilet in a home, but I have seen them in a few. If a toilet has a floor-mounted bowl, it should be fairly easy to replace. On those rare occasions when you have a true wall-hung toilet to replace, consult your plumbing contractor. Rerouting pipe to accommodate a floor-mounted model can be tricky and expensive.

A toilet is the easiest bathroom fixture to replace. An experienced plumber can complete the job, start to finish, in less than an hour. Some attractive toilets can be bought at very low prices, so this is an improvement that doesn't have to cost a small fortune.

It takes a lot of work to replace showers and bathtubs because the job involves not only plumbing work, but wall repairs. You might be better off having an existing bathtub or shower refinished. The companies that refinish plumbing fixtures usually do good work and their prices are very attractive when compared to a full replacement. Most bathrooms are limited to three fixtures—a toilet, a lavatory, and a bathing unit. In cosmetic conversions, each of these fixtures should be assessed for refinishing or replacement. If there's sufficient room, sneak a whirlpool tub into the picture (Fig. 15-9).

Electrical fixtures

Electrical fixtures in older bathrooms usually leave a lot to be desired. Most older bathrooms have very poor lighting. As a cosmetic contractor, you might get by with replacing existing light fixtures, but the addition of new lighting might be in order. A favorite of mine is an oak strip holding four designer light bulbs. This type of light can be used as a straight replacement, and it adds so much more to a bathroom than a dull, plastic-shaded fluorescent light does. Strip lights are, of course, mounted on walls, so ceiling lighting remains a concern. Consider installing a light-fan, or heat-light combination unit in the ceiling. Some customers might ask for a fancy fixture, but a recessed light-fan combination is pretty standard procedure.

Windows and doors

A bathroom might have no windows and only one door separating the bathroom from living space. Tub and shower doors can be considered doors, and these are a nice improvement over mildewed shower curtains. If a bathing unit doesn't have glass doors, give a lot of thought to making this improvement. The few hundred dollars spent will blossom into a much richer bathroom.

As for windows, you might be out of luck. Not all bathrooms are built to have windows. If there is an existing window, you might be able to improve upon it. When a bathroom has an outside wall and suffers from a lack of natural light, you can use frosted blocks or fixed

15-9 *This whirlpool tub, with the surrounding tile, large windows, and unusual walls, makes for a grand bathroom fixture.* Pella

panels of glass to create something like transom lighting high on a wall. A similar procedure was discussed for kitchens.

The master bathroom in my home is fairly large. It contains a shower, a whirlpool tub, a large vanity and top, and a toilet. A walk-in closet adjoins the bathroom. To get extra light to the whirlpool, for reading while relaxing, I had my carpenters install a piece of fixed glass the width of the tub and about a foot tall. The glass is high enough on the wall that no one can see in, but light floods the whirlpool. Such a piece of glass costs so little that it's hardly worth mentioning, but the result is rewarding.

Octagonal windows work well in small bathrooms. Awning windows offer both ventilation and privacy. If you have an opportunity to install a window in a bathroom that doesn't have one, do it (Fig. 15-10). When working with a bathroom that has only a small window, replace it with a larger one. Can you put a skylight in the ceiling? If so, the room will get plenty of light. Use a skylight that can be opened, and the homeowners will not have as many moisture problems when they use the skylight as a vent. Light is an important factor in bathrooms. If you can't provide it with glass, make sure you have your electrician make up for the deficiency.

Accessories

A lot of contractors overlook bathroom accessories. You shouldn't. Accessories can have a lot to do with the success of a cosmetic conversion on a bathroom. Don't believe me? Well, let's look at some comparisons. If you walk into a bathroom and see square, chrome-plated towel racks, what do you think? Maybe nothing, but you probably think about how cheap the accessories are. Any contractor knows junk accessories on sight. I'm not saying that the towel rack isn't functional, but it's about the least expensive towel holder a builder or remodeler can install, short of a 16d nail.

15-10 *Notice the window, glass blocks, and skylight in this bathroom. Keep the design in mind.* Mannington

Bathrooms

Now imagine the same bathroom with a towel bar made of oak, and an oak towel ring hanging on the opposite wall. What do you think now? Are you a little more inclined to think that the property owner has better taste than the homeowner with the chrome-plated towel bar? We shouldn't judge people by their towel bars, but a lot of people do. It's true that the quality of bathroom accessories can make or break a cosmetic remodeling job.

What constitutes an accessory? Mirrors can be considered accessories, although their role in a bathroom can be very important (Fig. 15-11). Holders for toilet tissue are accessories, as are towel and facecloth holders (Fig. 15-12). Toothbrush holders, soap dishes, and similar items are accessories. They can be finished in wood, brass, chrome, or even gold. I prefer a natural wood finish.

How many times have you remodeled a bathroom and had your helper install accessories that you bought as a prepackaged kit for less than $20? I know a lot of builders and remodelers, myself included at one time or another, who have done this. My experience suggests this is a poor procedure. I'm not saying that you or your customer should invest hundreds of dollars in accessories, but I think it's fair to say that a new bathroom deserves more than the one-kit-fits-all type of accessories. Customize the bathrooms you remodel with good-looking accessories.

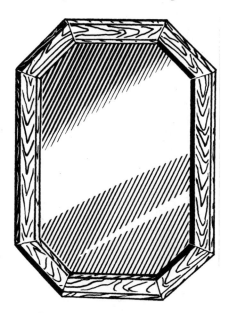

15-11 *An affordable, oak-frame mirror is a good choice for most bathrooms.* Nutone

15-12 *Towel holder.* Nutone

Medicine cabinets are a common bathroom accessory. Some have mirrors, and others have lighted mirrors (Figs. 15-13 and 15-14). Medicine cabinet doors can have wood finishes (Fig. 15-15), and there are even cabinets designed to hide behind wallpaper (Fig. 15-16). Invest a little money and time to give your customers the custom look they deserve.

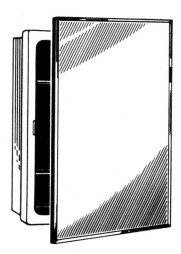

15-13 *Standard, mirror-front medicine cabinet.* Nutone

15-14 *Double-door, mirror-front, lighted medicine cabinet.* Nutone

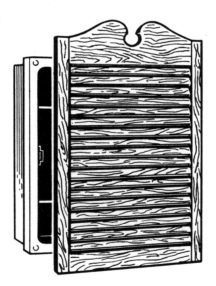

15-15 *Sculpted wood-front medicine cabinet.* Nutone

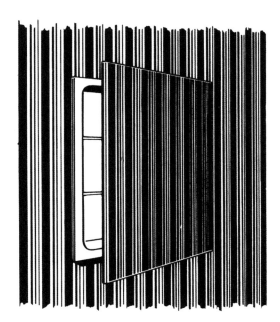

15-16 *Hideaway medicine cabinet.* Nutone

16

Living rooms

When I think of living rooms, I think of two types, formal and informal. I've never owned a home with a formal living room, but I've remodeled many homes that had such rooms. Some people and lifestyles call out for formal living rooms. People might buy a home and want a formal living room added, or an informal living room converted to a formal one. Chances are good that a remodeling contractor will work with a formal living room.

Since the focus of this book is on cosmetic changes, I will not venture into the subject of building additions for formal space. Instead, I'll concentrate on converting informal space to a formal living area and on revitalizing existing formal living rooms.

What makes a formal living room formal? This is a tricky question. When I think of a formal living room, I see expensive carpets and rugs, fancy window coverings, a lot of wood trim, uncomfortable chairs, white walls, and punched-tin ceilings. Of course, not all formal living rooms contain these elements. In fact, many formal living rooms are very different from my vision.

I don't know the true definition of a formal living room. In my opinion, a house that has a family room and a living room could be considered to have a formal living room, depending upon how it is appointed. I suppose a house doesn't need a family room to have a formal living room, but it seems more natural to me for a house to have one room for informal gatherings, children, and relaxation. This brings me to my personal definition of a formal living room.

I've never been one to entertain or travel in social circles. When people come to my home, they are people I know well and who probably would be uncomfortable in a formal setting. If I were to seat my guests in a sterile room with straight-back chairs surrounded by busts of ancient Romans, they probably wouldn't stay long. Visitors at my home seem comfortable in recliners, soft sofas, and rocking chairs. This is one group of people who come to my home, but another type of person visits.

My office also is in my home. Even when I've maintained commercial office space, I've always kept an office at home. The office, where I meet professional contacts, is separated from the other living space and is set up like a real office. It's not a desk and chair stuck in a laundry room, although that is exactly how I was set up when I started writing. I suppose that if I didn't have an in-home office, I might consider the need for a formal living room.

There are those who have jobs that make entertaining a part of their lives. Others just like to get dressed up and have people over. These people have more need for formal living rooms than I do. Based on my experience as a contractor, people who have formal living rooms generally know, with some certainty, what they want done during a cosmetic conversion.

When you get into cosmetic work involving formal space, you are likely to find yourself dealing with materials with which you might not normally work. Instead of pine woodwork, you might be using mahogany. A painted drywall ceiling might be used, but an ornate ceiling made of metalwork could be in order. Hand-painted wallpaper might be selected instead of paper bought out of a bin in a paint store. Some hardware for doors in formal space can cost more than several standard interior doors. The cost of materials will depend on how formal the formal space is.

A living room doesn't have to be decked out in high-dollar fashions to be considered a formal room. As I said earlier, most people consider any living room that is used primarily for entertaining guests to be a formal living room. Painted walls and standard carpeting are acceptable in a formal living room. Unless you are fixing up a room to look like some elegant living room in a mansion on a television show, you don't have to get into extremely expensive materials.

Three types

On the whole, living rooms tend to fall into three general categories. The first is what I'll call "average," the second is somewhat more formal, and the third is the all-out fancy style.

Average formal living room

An average formal living room usually is built in much the same way as a family room or standard living room. The room's furnishings generally are what designate it as a formal space (Fig. 16-1). Some building materials and trimwork are likely to be upgraded from what would be considered average, but most of the room construction is

Living rooms

16-1 *A formal living room.* Pella

typical of modern building practices. Let's take a quick look at some of the elements of an average formal living room.

Carpeting frequently is the flooring found in an average formal living room. The grade of the floor covering and pad is likely to be a little higher than the flooring in other parts of the home, but the materials usually are stock items. Walls normally are painted drywall, although some will have wallpaper or be half wallpaper and half paint. Electrical devices, such as switches and outlets, have normal, plastic covers. The ceiling is painted or textured drywall. Ceiling lights are uncommon. There usually are no doors separating the living room from the remaining space in the house. Woodwork normally is colonial trim and casing, and it's possible that crown molding or chair rail is in place. Aside from a few upgrades, the room will not be very different from other rooms in the home.

A step above

Some formal living rooms will be a step above average. These rooms might have hardwood floors and crown molding, perhaps with wainscotting and wallpaper covering the walls. The ceiling probably will be the same as you would find in an average formal living room. Electrical devices will have fancy covers, and woodwork will be ornate. This type of formal living room stands out from other parts of the home.

Fancy formal living rooms

Fancy formal living rooms can consist of so many options that I probably couldn't think of them all (Fig. 16-2). To start with, the floors normally will be wood, with expensive rugs strategically placed. Walls might be painted drywall, but are more likely to be covered with wainscotting and wallpaper. Unlike other formal living rooms, where overhead lighting is nonexistent, fancy living rooms might have high ceilings adorned with chandeliers or similar types of ceiling fixtures. Woodwork will be borderline excessive and exquisite. French doors, or equally impressive doors, will create a barrier between the formal living room and the rest of the home. Electrical devices will be dressed up with designer covers, and all other aspects of the room will be pushed to social extremes.

Selecting a style

With three types of formal living rooms from which to choose, you and your customers must decide which type of room to create. Part of the decision will be made for you by the type of house with which you are working. A small tract house with a total of 1400 square feet probably isn't a candidate for a formal living room, and especially not for a fancy one. A colonial home, with 2600 square feet on an acre lot, could be a candidate for a fancy formal living room. You have to match the room to the house and to the homeowner's budget and circumstances.

I feel that much of the decision on how far to go with a formal living room depends on the homeowners. If your customers want a nice, clean room in which to entertain occasional guests, an average

16-2 *A fancy formal living room.* <small>Thibaut, Lis King</small>

formal living room should fill the bill. People who need a fancy formal living room probably will tell you what they want. I think the hardest decision, in most cases, is deciding between an average formal living room and one that is a step above average. In my opinion, contractors normally don't have a lot of opportunity for voicing an opinion when it comes to fancy rooms.

If I'm right in my assumption, we should focus on average formal living rooms and those rooms that are just a little better than average. Deciding between carpeting and hardwood flooring can be tough. A customer who would like a French door installed might agonize over the cost of installing something that isn't actually needed. The same applies to wainscotting, crown molding, and other common elements of formal space. Let's start with a homeowner who has just purchased a home and wants you to convert an existing living room to a formal living room.

Our subject house is a split-level home. Most of the lower level is unfinished. People enter through the front door into a foyer, and have the option of going up a few steps or down a few steps. The upper level contains a living room, a kitchen, three bedrooms, and a bathroom. As you stand in the foyer, you can see through an open railing, right into the living room. A wall screens the kitchen from the foyer, but the kitchen can be seen from parts of the living room. A hall leads out of the living room to the bedrooms. The downstairs has a finished room visible from the foyer and a considerable amount of unfinished space that is concealed behind walls and doors. How would you go about converting this house to give it a formal living room?

The job you've just been given is a pretty easy one. The house and its layout are well-suited for a cosmetic conversion. Since the homeowners have young children, they need a place for the kids to play and for the family to get together. The unfinished space downstairs can solve that problem. Convert it to a family room, and the living room need not be used for anything other than formal meetings. One problem with the room, from a formal standpoint, is the view of the kitchen. You have to eliminate this, and there are a few ways to do so.

I said previously that I favor pocket doors for hiding a kitchen. This could be an option for our subject house. If you don't want to go to the trouble and expense of a pocket door, have your customer consider using a tri-fold screen to block any view of the kitchen. These screens add to the decor of a room and are efficient at hiding all sorts of things. Many of the screens carry Oriental markings, but others are available. Something as simple as standing a screen up in the room can produce a formal effect.

The hall leading out of the living room could be considered a distraction. You could install a door in the hall entrance to close it off, but this might make the house feel too small. The best course of action might be to dress up the hallway and leave it visible. Guests might have to visit the bathroom, so you might as well commit to making sure that the hallway complements the formal living room.

People entering our subject house start in the foyer. If this area has average flooring, I would upgrade to something better, such as quarry tile. Assuming that the steps are carpeted, I probably would replace the carpeting. You might consider replacing the carpeted steps with finished wood steps, since there are only a few, but I don't think I would do that unless I were installing wood floors in the living room. To keep costs down and to maintain an average formal living room, I would simply update and upgrade the carpeting on the steps, in the living room, and in the hallway. Remember, the hall is essentially a part of the living room, so it must maintain the level of formality presented in the living room.

Having decided on what to do with the flooring, I would look to the walls. The existing walls are painted drywall. I either would clean and repaint them, or consider installing wallpaper. Considering the design of the house, its size, and other factors involved, I probably would suggest that the homeowners stick with painted walls and ceilings. After all, they are spending money to convert the downstairs into a family room, so money might be a consideration. Depending on the layout of the living room, I might consider installing a two-panel, metal-insulated, floor-to-ceiling glass door in the back wall of the room, opening on a small deck or terrace. This would make a tremendous difference in the room's appearance, and would add an element of distinction. It also would be functional, providing easy access to the outside.

In this particular house, I wouldn't worry about crown molding or other ornate trim. As long as the baseboard and casing trim is attractive, I'd leave it alone, except for giving it a new coat of paint or stain. This type of living room can be converted to a formal space for very little cost. Now let's look at a different house and see how the circumstances change.

Our second house is a large Cape Cod. It has a room designed to be a formal living room but never used as one. The previous owner used the room as an art studio. Visitors entering through the front door find themselves in a foyer with steps in front of them, a room to their left, and a hallway to their right. The room on the left is the formal living room. It's ideal, in that it is separated from the rest of the house and is convenient to the main door. Guests have only to enter the door, hang up their garments, and step into the formal living room. There is, however, a problem. The proposed formal living room is surrounded with dark wood paneling. Dents mark the ceiling, signs of abuse from tall objects. The wood floor has turned black from years of neglect and water spills. The person using this room as a studio had no respect for its intended use. It's your job to fix it. How are you going to do it?

Things aren't as bad as they might seem. Fortunately, this room has all the characteristics of a formal living room, and even has the potential to be a step above average. The room looks rough, but there are a lot of fundamental ingredients with which to work. The dented and scarred ceiling can be hidden with a good texturing job. Refinishing the wood floors will make them look new. As for the walls, the paneling has to go, but this is no big deal. Rip it out, add some wainscotting, a little drywall, and some wallpaper, and you've got the makings of a nice formal living room. As you can tell from the two examples, the circumstances under which you work with conversions for formal living rooms can be diverse.

It's obviously more of a challenge to create a formal living room than it is to refurbish one. If you are starting with a formal room, the cosmetic additions can be similar to those of any other room in a home. You might have to do little more than clean or replace flooring and add new paint. Chances are, though, that you will get into more cosmetic work with a formal room than you would with a standard room. This is partially because many formal rooms have more in them to begin with.

As houses age, the construction components begin to date them. This isn't necessarily a bad thing. Some old woodwork is more attractive than new wood. Wallpaper is an element of formal rooms that often ages poorly. The paper can become discolored or tattered, and its pattern can scream out that the room is old and in need of help. Fortunately, replacing existing wallpaper is not an unusually difficult job. The expense can add up, but there are low-cost options that look good.

Many of the homes that I've worked in have been old. A lot of them have had tall ceilings and many have had tin ceilings. When you get into older homes like these, you can do more damage than good by replacing old things with new materials. It's possible to destroy the charm of the room by replacing too many original materials with new materials.

If you've worked with old houses, you know the look. There's the tin ceiling, wallpaper, and some type of wood floor, often darkened with age. The trim around doors has little squares at each top corner. These squares are a good example of where you can go wrong by getting too modern. If you have to replace the trim, consider replacing it with modern replicas of what is already there. Some carpenters are set up to make their own trim materials, but most contractors buy their materials. This isn't a problem. Several companies manufacture modern trim pieces that look for all the world like the old pieces they are meant to replace. Keep this in mind if you are restoring an existing formal space that benefits from its age.

Material selection

Let's talk about the basics involved in selecting materials for formal living rooms. When you are dealing with an existing formal space, there is a good chance the flooring will be some type of finished wood. Wood floors are durable, but they get cut, scratched, and discolored. This usually isn't much of a problem. Any good refinisher can bring this type of floor back to life. With some sanding and sealing, an old floor can look new again, and the cost is much lower than that of a full replacement.

If you find a formal room that is covered with old carpeting, don't be surprised to find wood floors under the carpeting. When wall-to-wall carpet became fashionable, a lot of people had it installed. A good many of these installations took place over existing wood floors. I've remodeled a number of houses in which this was the case. When the carpeting is removed, the flooring under it can look like a total loss. It usually isn't. Again, a refinisher can bring out the best in wood floors. I can't remember the number of customers I've had who hired me to replace their carpeting and then were delighted to find original wood floors had been hiding under their old carpeting. In my opinion, wood floors should be preserved, whenever possible, for formal living rooms.

Most newer houses are likely to have carpeted floors. Some will, of course, have wood floors beneath, but the trend for many years has been to install carpeting directly over subflooring. This isn't a problem. You don't have to tear out the carpeting and install wood floors to make a living room formal. It is desirable, however, to use a high grade of carpeting and pad in formal spaces.

As you work your way up to walls, you might find that you are dealing with plaster in older homes. Some plaster walls remain in good shape. The ones that aren't are something of a pain to work with. A few contractors go to the trouble of floating out old plaster walls and making them good again. Most, however, rip out the plaster and replace it with drywall. I've tried both approaches and feel that the rip-out method is usually best.

When you tear out plaster walls, you have to be prepared for poor working surfaces. The studs you'll find under plaster rarely, if ever, can be used for hanging drywall without planing and shimming. Most of the wood used with plaster was never planed. Trying to get a straight wall out of the studs found in most plaster jobs is impossible, unless you take the time to furr out the wall. I learned this the hard way. My first plaster rip-out went well during the demolition phase. It was not until I hung several sheets of new drywall that I realized just how far out of plumb the walls were. Take my word for it, you will have less frustration if you just go ahead and fix the walls before you start hanging new drywall.

For wall coverings, you can choose among paint, wallpaper, or wood. A combination of these materials is acceptable. Paint is common and accepted by most people. Wallpaper is a little more formal and will be what it takes to satisfy some customers. Wood, such as wainscotting, is optional and certainly isn't needed. There are people, however, who feel a room just isn't formal if it doesn't have wainscotting. This is something for you and your customers to discuss.

Ceilings in formal living rooms can be done with nothing more than paint. A textured ceiling looks fine in a formal living room, but I prefer a flat, painted ceiling over a textured one in this type of setting. Stay away from ceiling tiles in formal rooms, but consider some of the alternative ceiling applications available. There are a number of choices, ranging from fake planking to punched tin.

Windows can say a lot about a room. If you are working in a room in which standard, double-hung windows dot the walls, consider replacing them with something more stylish. Tall windows with rounded tops can make quite an impression on visitors, and there are several other types of windows that will give a room a formal look. The same is true of doors. Break away from common materials and look for units that add a little spice to the space.

I think the most important thing to consider when doing a formal conversion is the fundamentals. Try to separate the room, with doors or design, from the rest of the house. Keep the materials used at a level above the rest of the house. By making a room formal, you are setting it apart from the rest of the house. In the simplest of terms, this is your goal. Follow the suggestions in this chapter and the wishes of your customers. Pay attention to detail and make the room look special. If you do these things, you are well on your way to creating fantastic formal space. Remember, formal doesn't have to be stuffy. Contemporary living rooms can have a lot of appeal (Fig. 16-3). The real reward will be seeing the room once it has been decorated.

16-3 *A contemporary formal living room.* Velux

17

Family rooms

A family room can be a fun place, the area of a home in which families gather for some of their best times together. These rooms see so much use that they often wear out before the rest of a home, setting the stage for cosmetic contractors to do their thing. Fixing up run-down family rooms can become something of a specialty. If you become known for your good, creative work with family rooms, there is a good chance you can get return business for years to come.

Not all houses have family rooms. Many homes have one room that doubles as both a living room and family room. When this is the case, the need for cosmetic improvements is more apt to keep you busy. People might look the other way when a family room gets a little tired, but if the room also serves as the living room, they probably won't wait very long to have cosmetic work done.

During my remodeling career, I've informally kept score of what types of jobs I do the most and which jobs yield the highest returns. Bathrooms are my top producer, both in volume and profits. Kitchens are in second place, followed by family rooms. My records might not mirror the results of other contractors, but a lot of remodelers with whom I've compared notes agree with my statistics. I suspect that contractors all over the country would agree, at least in general, with my findings. There are exceptions, of course, but we are talking in general terms.

Some contractors might make enough money to take a long vacation after a kitchen job. Others find their rewards in designer master suites. For me, it's bathrooms, kitchens, and family rooms. There is a method to my madness. Consider the three rooms I'm talking about. Can you think of any other rooms in a home that see more use and activity? And all that activity makes these rooms good candidates for remodeling. The fact that these rooms are ones that receive the most public attention doesn't hurt. It's not often that visitors tour bedrooms, sewing rooms, and laundry rooms. But they often ask to use the bathroom, and many casual visitors enter homes through kitchens. Family rooms and living rooms that serve double duty also attract visitors. I think this is why family room work is an excellent opportunity for cosmetic contractors.

The preceding chapter dealt with three basic types of formal living rooms. When it comes to family rooms, there are so many variations possible that I won't begin to list them. You could have a room that houses a billiard table, entertainment center, and comfortable seating arrangements. The next family room you work on could resemble a semiformal living room, or you might find a family room that doubles as a music room (Fig. 17-1). Family rooms can have brick walls, beamed ceilings, decorative paneling, or just about anything else imaginable. This presents remodelers with pros and cons. You are nearly unlimited in what can work in a family room, but trying to reach a decision on what to do can be a chore.

For the sake of discussion, it might help to break family rooms down into a few categories, with the first category being family rooms that also serve as living rooms (Fig. 17-2). This type of room will be our most formal entry in the list. A second category is family rooms that are separate from the living room. A third category is basement family rooms, because they are common and present some unique problems to remodelers. By using these three types of rooms as a baseline, we should be able to define our options more clearly.

17-1 *A musical theme is the subject of this family room.* Environmental Graphics, Inc., Lis King

17-2 *A combination family room and living room.* Velux

Basement family rooms

Let's begin with basement family rooms. If you work in an area where basements are the norm, you probably have converted more than a few of them into living space. In doing so, I expect you have created a number of family rooms. Expanding families often turn to their basements as options for expansion.

Converting basements into living space is a way of getting usable space at the lowest possible cost. The investment isn't normally considered a good one in terms of appraised value of the improvement, but it is a sensible way to expand a house. The big advantage to basement conversions is the low cost per square foot. When people make these conversions, they often use moderate- to low-grade materials. Since the room is in a basement, the homeowners are not usually compelled to go with top-of-the-line products. These same people might cut some corners when first doing a conversion. If money is tight, they might do as little as possible to create a clean, safe playroom for their kids.

As time passes, some folks find that they enjoy their basement family rooms so much that they want to make them more comfortable. If economic conditions have improved, a couple might decide that the cheap carpet they installed originally should be replaced with a more luxurious type.

Many homeowners do their own family room conversions in basements. It's not unusual for them to go to the local lumber department store and stock up on paneling that's on sale. For do-it-yourselfers, the paneling seems like a good deal. They don't have to tape and sand joints on drywall, and the paneling will not need to be painted. What they don't realize is that paneling in a basement often makes a room very, very dark. They have to live with their mistake for awhile, but sooner or later, they probably will have a contractor replace the paneling.

Ceilings in basement family rooms can be interesting. Sometimes they are nonexistent, with the floor joists of the first level of upstairs living space all that can be seen of the ceiling. In other cases, ceiling tiles have been installed. The decision to use these tiles sometimes is based on a need to hide pipes or ducts. In general terms, some basement family rooms are a contractor's challenge, to say the least. This, however, is not always the case.

I've done more basement family rooms than I can remember. Some have been very nice, indeed. The mere fact that a family room is in a basement doesn't mean it is some thrown-together junk room. Rooms in basements, however, can force cosmetic contractors to work around some unique problems. Lighting often is a concern. Moisture and mildew can make the cleanup of a basement room difficult. Getting natural light into many basement rooms just isn't feasible. Support columns frequently rob basement rooms of usable space. Even the heating or cooling systems used with basement rooms can present problems. You have to be prepared for such obstacles.

Flooring

Flooring selections for basements depend on many factors. Most flooring is applied to the concrete floor that exists in a basement. Many types of carpet can be used for this type of installation, and some people prefer sheet vinyl. I don't like to install vinyl on concrete floors. The walking surface is hard and is usually cold. Carpeting, however, can suffer from moisture problems. What's the answer?

I use carpeting in most of my basement work. If the basement is damp, I install a drainage system to control water infiltration before putting down any flooring. Dehumidifiers can be used, if needed, to control excess moisture. Carpeting softens the walking surface and adds to the warmth of a basement room.

On occasion, I've had my crews build a subfloor structure over a concrete floor. This is easy to do. You lay pressure-treated lumber on the concrete and shoot it into place with a powder-actuated nailing tool. Plywood then can be installed on the lumber, providing a normal subfloor with which to work. At this point, you can use any type of flooring. This type of system is used when wood flooring is to be installed, it works well for vinyl flooring, and it can be used with carpeting.

When I think of family rooms, I normally envision carpeting on the floor. To me, this is the most user-friendly flooring available for a family room. However, you should not become locked into carpeting. Parents might want vinyl flooring so that their children can race cars across the floor or do other things that only a smooth, slick floor will allow. Another viable reason for vinyl flooring is that it's easier to clean than carpeting, and this can be a plus for parents with young children.

Walls

I think walls in family rooms should be tough. If kids want to bounce a ball off a wall, what better room to do it in than a basement family room? When doing cosmetic conversions, I've often installed wood paneling half way up family room walls. By keeping the paneling on the low portion of the wall and using drywall with a light-colored paint on the upper section, the room does not suffer from being too dark.

Drywall is fine for walls in family rooms, as is light paneling. The key to wall coverings in basement family rooms is to keep them light in color. Any dark colors, whether paint or paneling, can make a basement room dreary and dark. This just won't do.

Some people like to set a theme for their family rooms. I've done rooms with both paneling and murals in which specific themes were represented. While I've done a lot of paneling and some murals in family rooms, I can't remember ever covering the walls of a family room with wallpaper. Paint, based on my experience, is the most common wall covering in family rooms.

Ceilings

Ceilings in family rooms range from painted drywall to acoustic tiles. The tiles offer a few advantages in basements, a big one being that lowering the grids for tiles results in exposed wiring and piping being hidden. This saves a lot of time and work compared to moving all obstacles into joist bays to allow for a drywall ceiling. Perhaps one of the most overlooked advantages to acoustic tiles in basement family rooms is that they reduce noise. Kids yelling and playing in the basement will not be nearly as disturbing to people in the room above them if an acoustic ceiling is used. If your customer really wants to deaden the sound, you can add fiberglass insulation between the ceiling tiles and the floor to the room above.

Lighting

Lighting may be one of the most important aspects of a basement family room. This is especially true when the basement is devoid of full-size windows. Lamps just aren't good enough when you need to light a basement room that doesn't have the benefit of windows or glass doors. Walk-out basements and half-basements aren't so bad, but buried basements are awfully dark.

Track lighting is my favorite solution to lighting problems in basement family rooms. This type of lighting is not too expensive, and it's pretty easy to install. Regular ceiling lights also can be used, but I like the look of track lighting, and you get more light that can be directed to where it's needed. Now, let's move upstairs and check into the possibilities for main-floor family rooms.

Main-floor family rooms

Main-floor rooms that serve solely as family rooms can be approached in ways similar to those described for basement family rooms. Due to location, there are some differences in the two types of rooms. Main-floor rooms are easier to work with than basement rooms. Following is a sample scenario on this topic.

Assume that you are working in a two-story home that has a formal living room and a separate family room, which was built years ago as an addition to the home. The room has a stone fireplace, and the existing walls and vaulted ceiling are painted drywall. The finish flooring in the room is carpet. Your customers don't want to undertake major work, but they would like to see a significant change in the room. What can you do?

The fact that the room has a stone fireplace tells me it could benefit from a rustic motif. Vaulted ceilings are meant to be accented with exposed beams, and the beams fit the rustic look. Already we are on the path to a new family room. Build some false beams, or buy prefab models, for the ceiling. Have your electrician install a ceiling fan with a light kit. Maybe the carpeting can be cleaned well enough that it can stay in place. If not, replace it. The walls are all that is left to figure out. Maybe some weathered barn boards should be installed as wainscotting. If money is too tight, a new paint job will do. Ideally, the room would benefit from some type of wood to tie in with the beams, whether it be the barn boards, some rough-sawn siding, or some T&G planking. Install new wood trim, a primitive type, and stain it. With a few wall hangings and decorations, the room should not even resemble its former self. These types of conversions are fast, fun, and affordable.

Combination rooms

Combination rooms, those that are used as both family and living rooms, can be the most troublesome when it comes to finding cosmetic answers. Because these are dual-use rooms, it can be difficult to pin down the types of materials and finishes to use in them. It usually isn't fitting to make an extreme statement in a combination room. An extremely rustic family room can be fun and appropriate, but rustic features might not work for a living room. By the same token, you can't use a formal living room for a family room and expect it to be comfortable and homey. The dilemma is finding the right mix.

If I had to guess, I'd say that a majority of homes have living rooms that double as family rooms. I'm certain that houses in the areas where I've lived fit this profile. When homeowners are forced to use one room for two purposes, they must make compromises. This is to be expected.

What should you do with a combination room? There are no rules for what can be done. It's up to the property owner to decide what will happen, but you should be ready to offer advice when needed. Painted walls are usually the best compromise when working with a combination room. Painted or textured ceilings are fine. Carpeting is the flooring of choice, and lighting should be good. Other elements of the room depend on the style of the home and the personality of your customer. You might add a wet bar (Fig. 17-3).

The homes I've built for myself have had contemporary flairs. I'm big on vaulted ceilings and ceiling fans. Six-panel doors and colonial

17-3 *The addition of a wet bar will please some customers.*

trim have been included in all of my personal homes, with the exception of one used home I bought. Skylights and textured ceilings have been in every home I've built for myself. I like exposed beams, but I've only used them in one of my own homes. My walls have normally been painted drywall, and my floors have always been covered with carpet. I guess you could build a case for me being a creature of habit when it comes to certain things. My preferences don't reflect the demands of all homeowners. Neither will yours. We have to put aside personal preferences when advising customers, and this can be hard to do.

Rooms that must be used for entertaining guests should be a little more refined than classic family rooms. The old recliner that your cat has used for a scratching post and your dog has used for an afternoon bed might look fine in a family room, but it probably isn't the image you want to display in a living room. A chair is furniture and doesn't fit into a contractor's requirements, but this offers a good example. When planning a cosmetic conversion for a combination room, you have to weigh all of the choices. If something is borderline, the option of making it appropriate for a living room usually will win out.

While I wouldn't normally put wallpaper in a family room, I might consider installing it in a combination room. A heavy theme might be appropriate for a separate family room, but it probably isn't a good idea in a living room. You must exercise judgment when sorting through remodeling options. Since combination rooms usually are living rooms first and family rooms second, you might have to attack the job in a different way than you would in a separate family room.

Combination rooms are tough to work with. You will have to talk with your customers and determine their priorities. For all you know, they might decide that they live in the house and use it more than anyone else, and thus want the combination room to first be a family room, and to be a living room second. You won't know until you discuss the issue, and you should discuss it carefully before you plan too much work.

Family rooms can be a lot of fun to work with, and they can provide years of enjoyment to families. These rooms might not be the most important ones in homes, but they rank high on the list. Devote some special effort to showing homeowners how to make the most of their family rooms.

18

Children's bedrooms

Children's bedrooms offer a wealth of opportunity for cosmetic contractors. As children grow, their needs and tastes change. A room that was fine for a 3-year-old isn't appropriate for a 7-year-old. Not only do growing children require changes in decorative aspects of their rooms, the need for different room appointments can vary from age to age. For example, a child might develop a need for bookcases or a computer station. Older children might want a window seat or some other type of built-in improvement.

Another reason why children's bedrooms are popular targets for cosmetic contractors is the fact that children can be rough on their rooms. Grape juice spilled on a carpet can leave a stain. Posters stuck on a wall can mean drywall repairs and painting in the future. All sorts of scuffs, scratches, dents, and other damage can occur in children's bedrooms.

I recently talked with a woman about a bedroom that was in need of refurbishing. Her daughter had outgrown the room's decorations, including wallpaper covered with teddy bears, and was embarrassed to have her friends over for visits. Structurally the bedroom was in pretty good shape, and the work required was all cosmetic. This is a prime example of the type of work you are likely to encounter. Since this job is such a good example, let me tell you what my assessment of the room was and what improvements were planned.

The room is assigned to a girl about to have her eighth birthday. The room is small, but not so small as to limit all options. A visual inspection of the room reveals that the girl is an avid reader. Books are everywhere. She also is a collector of many things, and her collections cover the top of her dresser and chest of drawers. Additional collectibles are kept on the closet shelf. Stuffed animals sit all around the room. This bedroom is navigable only on designated paths around the stuffed animals and other objects.

Opening the only closet in the bedroom, I find it overflowing with clothes and personal possessions. There is a definite need for some better organization in the room. A textured ceiling, in good shape, hung over worn and stained pink carpeting. The baseboard trim had been ravaged by a pet rabbit that once occupied the space. Tooth marks and splintered wood are evidence that the rabbit enjoyed chewing on wood.

One of the better features of this corner room, which has three windows, is the plentiful natural light. One window is on one wall, and two others are side by side on the other wall. The ceiling light is one of the old bug-catcher types with a clown and balloons painted on it. The only parts of the room that don't need attention are the ceiling, the windows, and the doors. Now that you know the layout of the job, what suggestions would you offer for improvements?

This job is typical of the type of work required in children's bedrooms. In order to make good suggestions for improvement, you have to get to know your customers and their children. For the sample room we're discussing, I had many ideas, a lot of them based on techniques I'd used in the past. Before offering my thoughts, however, I like to talk to the parents and children. This gives me more insight into what to suggest, and I have a better chance of offering long-lasting solutions. Now, what types of changes did you come up with?

There are many simple solutions to the walls and floor. New paint or new wallpaper can replace the teddy bears. Replacing the carpeting is no big deal, and that will take care of the flooring problems. But the storage and organizational needs require more thought and planning. There are three key issues to be dealt with. All of the stuffed animals need a new home. Books shouldn't be piled on the floor, and the girl's collectibles certainly shouldn't be stashed away in the closet. How will we solve these problems? It's really pretty easy.

One way to deal with the stuffed animals is to install hammock-type netting in which to store them. This is an inexpensive solution that works well. The netting is suspended above the living space, probably in a corner, and the stuffed toys are stored in the netting. I've used this approach many times, but it is not a part of my plans for the room we are discussing.

I mentioned that this room is small. With the bed, a chest of drawers, and a dresser, there isn't a lot of floor space to work with. My solution to this problem is a combination window box and bookcase. Again, this is something that I've suggested to many homeowners, and it has always produced good results. Let me tell you how the setup works.

My carpenters build a box along the entire length of a wall. In this case, it will be placed under the twin windows. The box can have pull-out drawers, but the one for this room will have three lift-up lids. All of the stuffed animals, and a lot of other stuff, can be stored in the box. This gets the goods out of sight, but keeps them accessible. Padding can be added to the box to make a window seat of sorts. Labor and material for this part of the project are minimal, and the results are fabulous.

At each end of the window box, my carpenters will build bookcases that are the width of the box and extend all the way to the ceiling. Even the tallest shelves can be reached by standing on the box. This combination unit takes up very little space and solves the problem of the stuffed animals and books. We are left with the collectibles.

To house the collectibles, I will have my carpenters build small display shelves on the sides of the bookcases. This will give the girl a place to keep all of her collectibles in plain sight, without worrying much about them getting broken. By using the window-seat arrangement for storage, the closet will be less crowded. Installing a few baskets and organizers will make the closet seem cavernous when compared to its former condition.

The window seat can be painted or stained. In the sample house, it will be stained. A couple of hundred bucks is all it takes to solve all of the space deficiencies in the room. With this done, the rest of the work involves straight construction issues, such as flooring, trim, and wall coverings.

As it turned out, the child who sleeps in the sample room is interested in astronomy. She wants a telescope for her birthday. Upon hearing this, I suggested a space mural for one of the walls in her bedroom.

(I should stop here and explain something. I meet with parents and children simultaneously. But when I make suggestions for improvements, I do it with only the parents present. There's not much worse that offering a suggestion that a child loves and a parent hates. It's up to you to conduct your business in a way that is comfortable for you, but I strongly suggest that you make recommendations to parents alone, before you run them past the kids.)

The girl said yellow is her favorite color. When I showed the customers wallpaper samples, I made sure to include some that contained yellow flowers. We also discussed yellow paint. The flowered wallpaper won the toss. Carpeting was the next item on the selection list. A neutral, beige color was picked. Then we had to do something with the Bozo light. I found a nice box-type light with a wood border and pebbled shade. The parents loved it. Replacing the trim will be done with colonial woodwork that will be painted. I think the finished room will look great, and it will be generic enough to last the little girl for many years to come.

Flooring

The flooring in a child's bedroom normally will be carpet. This a sensible, affordable flooring that gives good overall surface. One thing you might want to suggest is that your customers invest in a good stain-resistant carpeting, especially if the children are young. It doesn't hurt to install carpeting that will not show dirt and stains easily. For example, a multicolored brown carpet would be more practical than a white one.

There tend to be two schools of thought about installing carpeting in a kid's room. The first calls for buying the best stain-resistant carpet available. The second is for the purchase of cheap carpeting, on the assumption that it will be replaced frequently. Either way works, depending upon the desires of a customer.

You could suggest other types of floor coverings for a bedroom, but I don't know of any that would be more appropriate than carpeting. Viny flooring might be desirable, due to its durability with spills (Fig. 18-1). Wood floors would be the only other type I would consider, and I don't think that they are a great choice for a youngster's bedroom.

Ceilings

Ceilings in children's bedrooms can be simple or a bit extravagant. Most will be painted or textured drywall. This is fine. In some cases, especially in rooms occupied by infants and very young children, ceilings with painted designs are nice. On many occasions, my crews have taken a white ceiling and added a lot of color and designs to it. This gives a child a lot to look at, and the room becomes special when it has some type of theme painted on the ceiling. It could be circus scene, a representation of the solar system, or just about anything else.

The problem with customizing a ceiling for a very young child is that the child might outgrow the ceiling. Few teenagers are going to accept a ceiling full of nursery-rhyme characters. This is something

Children's bedrooms

18-1 *A child's room with vinyl flooring is easier to clean up after spills occur.* Congoleum Corp.

to keep in mind when advising parents, but you will find that some parents are willing to make cosmetic changes every so often, and won't mind the cost of repainting the ceiling.

Walls

Walls provide a creative contractor with a host of possibilities. You can paint the walls white and then have designs drawn on them. I've seen rooms where the walls had building blocks and letters of the alphabet painted on them. One room had an elaborate painting of trains and associated items on the walls. Murals also work well, whether they depict a sports scene or a nature scene. Stencil borders are another possibility (Fig. 18-2).

Many rooms have plain, painted walls, and there is nothing wrong with this. Walls painted in a color that is not outrageous can give years of good service. But customized walls can make a child feel more secure and comfortable. This comes down to a parental decision, but you should offer some suggestions for various types of wall coverings.

I usually avoid wallpaper in rooms where young children will be active. It's hard to remove marks made with markers and crayons from some types of wallpaper. If you do suggest wallpaper, make sure the type you offer is washable and durable (Fig. 18-3). Kids can be tough on the walls of their rooms.

Built-in units

I've found that built-in units, whether they be computer stations, window seats, or closet organizers, are very popular with parents. Children

18-2 *Stencil borders can personalize a child's room.* Environmental Graphics, Inc., Lis King

Children's bedrooms 233

18-3 *Wallpaper can add flair to a room, but make sure it is washable and durable before you install it in a child's room.* Velux

accumulate a lot of stuff, and they need a place to put their possessions. Most parents prefer to have facilities that will allow the rooms of their children to remain neat and organized. This is difficult to do with a room that has nothing more than a standard closet.

Room size can inhibit your ability to do a lot with built-ins, but small rooms often are the ones most in need of help with storage problems. Many manufacturers offer all sorts of closet organizers. These are a good place to start when working with cluttered rooms. You probably will find that customers need more than just closet space.

Many children have their own television sets and VCRs. It's not uncommon for them to have stereo equipment, computers, and other electronic gear such as game machines. Parents can deal with the electronics by purchasing entertainment centers, but many stock entertainment centers are too large for small rooms. In those cases, customized built-in units are called for.

Each room offers its own challenges for built-in cabinets, shelves, and accessories. I've told you about the window seat setup with which I've had good luck. Let me give you a few other ideas. Assume that you have a room that is very small. The child who uses the room is a collector. All of the collectibles are small, but there are a lot of them. The bedroom has one outside wall and three interior walls, once of which is consumed by a closet. You have two interior walls to work with. What can you do?

The first idea that comes to mind probably is adding shelves to the existing walls. That could work, but the shelves might protrude too far into the living space, and could present a hazard as head-knockers. But suppose you open up the walls between the stud cavities, and build cased shelves for storage within the walls? This approach doesn't take up any existing living space, yet it provides a nice, recessed setting for collectibles. I've done this more than once, and people have always liked the effect.

If you choose to put shelves between wall studs, you could run into difficulty with mechanical systems. A wall might contain a plumbing pipe, heat duct, or electrical wires. These sometimes can be worked around, but there are occasions when mechanical equipment blocks a path beyond remedy. Be prepared for this possibility. Let your customers know that you could wind up having to patch the wall instead of installing shelves. In most cases, you will not run into many obstacles if you keep the shelves high enough to avoid wiring for electrical outlets.

I mentioned earlier the use of a hammock-type netting to hold stuffed animals. You can pull another trick out of your hat that works in a similar way. Put a pulley up near the ceiling. Use a mesh bag to hold light objects, like stuffed animals. When the toys are not in use, hoist the bag up to the ceiling and tie off the pulley rope. Kids can untie the rope and lower the bag when they want access to the contents. It's amazing how much stuff can be stored in this way with very little cost involved in the process.

Beds offer a good opportunity for built-in storage. Get rid of an existing bed frame. Build a foundation for the bed. Install pull-out drawers in the foundation. This gives easy access to items stored under the bed and increases a room's storage capacity without sacrificing living space.

When you start dealing with built-in units, there is almost no end to the possibilities. Room size and shape will have a lot to do with what you can and can't build. Creative contractors can always come up with some special additions that will improve the looks and function of a youngster's bedroom.

19

Adult bedrooms

Adult bedrooms must be approached differently than those of children. When you start doing cosmetic improvements on the bedrooms of adults, you must consider a lot of possibilities. Some of the basics are similar to those of other rooms, but there are other considerations to think about. For example, an adult bedroom can be finished with anything from a simple theme to a very complex one. Carpet on the floor plus painted walls and ceilings are all that a bedroom requires. With some creativity, however, you can personalize bedrooms.

The number of requests I've had for adult bedrooms has been lower than those for children's bedrooms. But some of the work done in adult bedrooms has been extensive. If we talk in terms of minimal cosmetic work, we must limit ourselves to paint, wallpaper, and flooring, for the most part. All of this can be a part of a cosmetic conversion in a bedroom, but the real difference usually comes when new features are added.

Fireplaces

A fireplace in a bedroom can be romantic. The second house I built for myself had a fireplace in the master bedroom. It was attractive when it wasn't being used, and it was cozy when a fire was flickering. As a remodeling contractor, I've added fireplaces to a few master bedrooms. Don't get me wrong, this is not the type of work that makes my phone ring off the hook, but it does come along every now and then.

Having a masonry fireplace built is a big job. It's also expensive. If you are adding a fireplace to a master suite strictly for appearance, you can consider using some type of prefab fireplace. Do you remember the free-standing fireplaces that were so popular years ago? I do. In fact, an orange one stands out in my mind. It was installed in a bedroom and took up so much floor space that I thought it was ridiculous.

The fireplace installed in my former bedroom was a prefab unit, but it was built into the wall. It didn't look like a masonry fireplace, but it looked good, and it worked well. One drawback was the size of the firebox, but I couldn't argue with the low cost of acquisition and installation. The fireplace wasn't used a lot, but it was a nice feature, and I enjoyed getting mesmerized by the flickering flames every now and then.

You can add a metal fireplace to some bedrooms very easily. The whole unit can be framed in and the framing can be covered with drywall. Triple-wall pipe can be used for the chimney. This type of setup isn't cheap, but it costs a lot less than a masonry unit would. I'm sure that most of your customers won't be in the market for a bedroom fireplace when they are looking for a change, but a few might be; don't overlook this possibility.

Terrace doors

If you are remodeling a bedroom that has an outside wall, consider installing terrace doors (Fig. 19-1). These doors can lead to a deck or balcony. Not only will the doors bring light into the bedroom, they will dramatically change the appearance of a room. Sliding-glass doors could be used instead of terrace doors, but I feel the terrace doors are a better choice. They tend to be more energy-efficient and their operation is often much smoother than that of sliding doors. Putting doors in a bedroom will make a major difference in how the room is perceived.

Skylights

If you have a bedroom with attic space above it, consider installing some skylights or roof windows (Fig. 19-2). Even if light boxes are needed, the skylights help bring in natural light. Depending on the location of a bedroom, you might consider installing roof windows for light and ventilation. Roof windows are not extremely expensive, and they make a huge difference in the appearance of a room.

Ceiling fans

Ceiling fans are a favorite bedroom add-on with me. It may be that I like fans, but it seems as if most of the customers I talk to also like them. Getting a fan equipped with a light kit can be tricky in bedrooms with standard ceilings, but you can have a standard fan installed. If there is enough headroom to allow for a light kit, go for it. In my opinion, a room can't have too much available lighting. Fans are good from a practical point of view, and they make good decorative items.

19-1 *Terrace doors used in a bedroom can be wonderful.* Marvin Windows & Doors

Built-in fish tank

Have you ever considered installing a built-in fish tank in a bedroom? It's fairly easy to do, and the visual result can be fantastic. The easiest way to do this job is to recess a standard aquarium into the wall of a closet. In the closet, the fish tank is accessible for cleaning and feeding purposes. From the bedroom, the tank looks like part of the wall. With some aquarium background paper, a

19-2 *Roof windows and skylights, especially ones that open, are a good addition to bedrooms.* Vellux

lighted hood, and some exotic fish, the effect is tremendous. Does this type of arrangement sound a little too far out for the types of customers you deal with? Maybe it is. Not everyone is into fish, but I've had fish tanks installed in the walls of about half a dozen bedrooms.

Windows

The importance of windows can't be overstated. If a room has dull, old windows, consider replacing them. Even if a room has fairly modern windows, you might be able to convince the homeowner to replace them. There are so many desirable window designs available that it can be extremely difficult to settle on a particular one. I like the look of windows that have an arched transom. I'm also partial to casement windows because of their full air-flow design. Windows are not one of the first construction components that I consider, but I never leave a job without evaluating them.

Ceilings

Ceilings in bedrooms can be textured, painted, mirrored, or beamed. Vaulted ceilings are especially nice. You can't always change a flat ceiling to a vaulted one, but sometimes you can. My crews have done this type of work on a number of jobs. After finding a job where the space above an existing bedroom ceiling was attic, I would arrange for the ceiling joists to be removed and have the ceiling converted to a vaulted look. Sometimes I'd add false beams for a better effect.

Most bedroom ceilings are plain. They might be textured or painted, but there usually isn't a lot about them to make anyone sit up and take notice. This doesn't have to be the case. Tongue-and-groove (T&G) planking makes a very attractive ceiling. If your customer wants a wood look without going to the expense of real wood, look-alike products are available. There isn't a lot you can do with a ceiling, but there are a number of product choices available to make a plain ceiling a little more noteworthy.

Floors

Bedroom floors tend to be covered with carpet. This is standard operating procedure and it's rare to find a typical house that has something other than carpet on the floor. That is not to say, however, that all bedrooms do or should have carpeting on the floors. There have been many jobs on which my crews tore out carpeting and replaced it with a different type of flooring. Contractors who work with old houses also have different conditions to consider.

I grew up in a town full of old houses. When I started remodeling, the houses I worked on were pretty old. A few of them didn't have indoor plumbing and many of them were so old that additions had been built to accommodate the bathrooms. It was common for these houses to have old plank flooring. The wood was often rough and ugly. I remember getting huge splinters in my feet when running around my grandmother's house.

The old plank flooring in a house needs to be dealt with before a new floor covering is installed. The quick, easy way of doing this is to add underlayment. Some remodelers cut out the planks and replace them with plywood. Either method can be effective. Removing the planks, or covering them with underlayment, is only the first step in creating a cosmetically appealing floor.

Newer homes don't have the old, individual floor boards. Plywood or waferboard is the usual subflooring found in newer homes. These materials usually are in good enough condition that nothing special has to be done to them in preparation for new flooring. A contractor need only remove existing carpeting and pad. The subflooring might have to be scraped to remove pieces of old padding, but the base usually is solid.

Most people prefer carpet in bedrooms. It's soft and warm to walk on. There are homeowners, however, who want their bedrooms to be a little different. This is especially true when a home has a master suite. Wood flooring is sometimes chosen for bedrooms. I've seen many bedrooms that had hardwood flooring, and I've used softwood flooring in some bedrooms. Tile can be used as a flooring in bedrooms, although it is not as popular as carpeting or wood.

The type of flooring used depends a lot on the style of the home. Spanish designs frequently make use of tile. Period Cape Cods use a lot of wood flooring. Contemporary homes might have wood, tile, or carpet. I remember a bedroom in one contemporary home that had a very impressive wood floor. It was made of T&G planking, but it was not the planking that made the floor unique. It was the pattern created with the wood that was special. The flooring was plain around the edges with a large star in the center. I'm not sure how the design was made, but the star was finished in a darker color than the rest of the flooring, and was very distinctive. The homeowner never explained what the emblem represented, but it was a floor I will not soon forget.

Most homeowners don't get into a lot of modification when they want cosmetic improvements made to their flooring. If they have wood floors, they generally have them refinished. When carpeting is the existing flooring, it usually is replaced with new carpeting. I have found that many customers upgrade to a better pad when they replace their carpeting. There's something enjoyable about your bare feet sinking into a plush carpet and pad.

I expect that most of your bedroom flooring work will be done with carpeting, but you should not limit yourself to this option. Give some thought to the particular house style and see if other types of flooring might work better than carpeting. The ultimate decision will, of course, be made by your customer, but it is up to you to make recommendations.

Walls

The walls of a bedroom offer the widest variety of cosmetic options. Many bedrooms exist with only painted walls; this is okay, but it's not normally exciting. The worst wall covering I can remember seeing in a bedroom was a fuzzy wallpaper that was red, blue, and foil colored. The fuzzy stuff was something else. I can't imagine waking up to those walls each day. But there are so many types of wallpaper and wall options from which to choose that there is truly something for everyone (Fig. 19-3).

19-3 *Wallpaper is usually a winner in adult bedrooms.* _{Thibaut, Lis King}

Paint is the most common type of wall covering. Standard bedrooms usually are painted in some neutral color. This is good for a spec house, but people sometimes like to customize their rooms. I've seen bedrooms with artistic designs painted on the walls. Graphic lines sometimes are used, and more picturesque scenes occasionally are painted on bedroom walls. When you think of paint, don't consider only a couple of coats of neutral paint. Give some thought to accents that might spice up the walls.

When it comes to wallpaper, there are so many patterns and colors that choosing one can be extremely difficult. I recommend staying away from the fuzzy stuff, but hey, your customer might like it. Wallpaper and stenciling can make a painted bedroom look much better. Since you are dealing with adult rooms, you have more sensible options than you would in a kid's room. The full array of wallpaper materials will be available to you.

Murals are more popular in rooms for children than for those of adults, but there are still grownups who like the look of murals. I think the key to success with murals is not to overdo it. Limit the mural to one wall, maybe two, but don't do the whole room with murals. Used in moderation, murals are a nice effect.

Doors

Adult bedrooms often have one entry door and doors for closets. Entry doors should be consistent with other doors in the home. Closet doors give you a chance to make a cosmetic difference. For example, consider installing closet doors that are mirrored on the exposed side. Not only can the mirror doors be handy when getting ready to leave home, they can visually increase the size of a room.

Closets with sliding doors might be more useful if the doors were replaced with bifold doors. When sliding doors are used, only one half of the closet is accessible at a time. Bifold doors solve this problem and give full access. The change in doors is more utilitarian than cosmetic, but it is a worthwhile consideration.

Built-in units

Built-in units for kids' rooms were discussed in the preceding chapter; that same concept applies to rooms for adults (Fig. 19-4). Many people like to have an entertainment center in the bedroom. It's not unusual for master bedrooms to have enough floor space to accommodate commercial units, but built-in units offer a few advantages. Since they are custom-built, they are the precise size needed. There is no wasted space. Stock storage units and entertainment centers are rarely as efficient as handmade units.

Built-in systems can be used for more than televisions and stereos. Depending upon the needs of your customers, you might build a small business center in a corner of the bedroom. This might be nothing more than a desk, or it might be a full-blown computer station. If your customer has designated a section of the room as a reading area, bookcases are a logical consideration. All sorts of storage can be accomplished with built-in units. And don't forget the possibilities of making closets more user-friendly by customizing them.

19-4 *Master suites with built-in storage are attractive.* _{WilsonArt}

Lighting

Lighting in most modern bedrooms is pretty pitiful. A person is lucky to have a ceiling light. Closets usually don't offer any lighting, unless they are walk-in closets, and even then, there is frequently a lone fixture in the ceiling of the closet. I feel that good lighting is essential in a house. Lights can serve more purposes than the need to see. Moods can be set with the right type of lighting.

Over the years, I've had a lot of recessed lighting installed in bedrooms. I've also used a fair amount of track lighting in bedrooms. Track lighting, which can be fitted with colored lamps that set a tone, is simple to install and offers a lot of versatility. Don't be blinded by the fact that a room has no ceiling light or only one light. Take some initiative and suggest new lighting arrangements.

Adult bedrooms generally are remodeled less frequently than other rooms in a home, but that doesn't mean they should be ignored. There is a lot that you can do with a bedroom when you take the time to consider various options. If you're going to be a successful cosmetic contractor, you need to become familiar with the materials available to you and offer them to your customers. A homeowner who calls you in to remodel a bathroom might be happy to have you put a new look on a bedroom, if you make the suggestion.

20

Basements

Basements can be full of cosmetic opportunities. Many basements are used, at least partially, as living space. In some cases, the space is used most often by residents and rarely by guests, but that is not always the case. Homeowners frequently convert their basements into family rooms and game rooms. Basements can be turned into apartments for teenagers or in-laws. If you get a call for a basement job, you can't be too sure of what you will run into until you get to the job site.

There are three types of basements to deal with, and each type offers opportunities and challenges. Buried basements are more common than daylight basements or walk-out basements. My definition of a *buried basement* is one that has tiny windows and either no access to the outside or access by a bulkhead door. This type of basement is the least valuable, but they undergo their share of conversion work.

Daylight basements have full-size windows, but no regular door for access to the outside. Builders sometimes outfit buried basements and daylight basements with stairs that lead up to a full-size door, but this doesn't qualify as a walk-out basement. The space in a daylight basement is more desirable than that in buried basements. Window size is the reason for this. When full-size windows are used, legal bedrooms can be built in a basement. This is not the case with buried basements.

Walk-out basements are the prime type for conversion projects. By having a full-size door that leads out of the basement at floor level, walk-out basements are considered normal living space. From a real estate appraiser's point of view, walk-out basements are the most valuable, with daylight basements coming in second. Buried basements are on the bottom of the list.

Basement upgrades

Some basements are completely unfinished when a cosmetic contractor is called in. Others have been finished into living space and need to be spruced up. There are variations between these two types of jobs. Some basements have bathrooms roughed-in, and others don't. But a bathroom can be installed in any basement with the use of a sump and sewage pump. You need a thick notebook to keep track of all the things you could do to make a basement more attractive and more versatile.

I recently reworked a basement family room to improve its appearance. During the job, I also remodeled a basement bedroom, a living room, and a bathroom. The living room and bathroom were upstairs. During the job, I replaced the front door to the house and added quarry tile to the foyer. Other odds and ends were taken care of. Most of the work, however, was in the basement. Since this job is fresh in my mind, I'll use it as our example.

The basement I worked in was a daylight basement, and contained a bedroom for a teenage boy. It was semi-finished, but in rough shape. There was no ceiling and the walls were covered with homeowner-hung drywall that had not been painted. The flooring was a remnant of carpet laid on the concrete floor. The room was usable, but not very nice.

The family room was in better shape than the bedroom, but it had problems. A very dark paneling had been installed in the family room and, despite two full-size windows, the room resembled a dungeon. A massive wood stove perched on a brick hearth took up a lot of floor space. Support columns sprouted from the loose carpeting and stretched to the ceiling, which was made of dingy acoustic tiles. The doors in the family room were cheap lauan doors that had begun to delaminate. The only overhead lighting in the room came from one fluorescent light that occupied a space the size of a ceiling tile. To put it bluntly, the room was not a pleasant place.

I was asked for advice on what to do with the rooms that were to be remodeled. My customers didn't know what they wanted, but they knew they wanted something different. The bedroom was easy. I suggested replacing the poorly hung drywall with new drywall. The boy was into baseball and bowling, so we used a mural on one wall that depicted a baseball field, complete with fans and players. I had my carpenters build a wall-mounted, glass-front trophy case for the teenager's bowling trophies. By moving a few electrical wires, we were able to install a normal drywall ceiling. Carpeting went on the floor, and a new light fixture was placed in the ceiling. Since the young man liked to read, my electrician installed a two-light, wall-mounted fixture near the bed location, to provide good lighting for reading in bed. We replaced the curtain that had served as a door with a six-panel, hollow-core door.

One problem we faced, which is common with daylight basements, was the ledge between the foundation walls and the stick-built walls. This house had a poured concrete foundation and 2 × 6 walls. We trimmed the ledge with one-inch trim boards and created a display area for the boy's collection of model cars. When the room was finished, you would not have recognized it from the pictures we took before we applied our cosmetic magic.

When we moved into the family room, we faced more problems. I wanted to get the support columns out of the room. To do this, we used a *flitch plate* to build a beam that could be hidden above the dropped ceiling. Since the old wood stove was no longer being used, we got rid of it and its hearth. Moving the stove and eliminating the columns made a huge difference in the available floor space.

The next thing to go was the dark paneling. It was ripped out and replaced with painted drywall. Trim boards were used on the ledge area. All of the trim installed was stained. Six-panel doors replaced the peeling lauan doors. New, commercial-grade carpeting replaced the old remnant that had been taped to the floor. Ceiling tiles were taken out and painted to look new. Track lighting was added in all four corners of the room, and an entertainment center was built along one wall. By the time the room was done, the homeowners were ecstatic. They couldn't believe how much larger the room appeared to be, not to mention how much better it looked.

The example we have just discussed is not uncommon. Many basements are finished into living space by well-meaning do-it-yourselfers who shouldn't be doing their own work. The quality of existing conditions in basements is often worse than what you would encounter in other parts of a home. For whatever reason, a lot of people figure they can be their own remodelers as long as they confine their efforts to a basement. Ironing out the wrinkles made by someone else isn't fun, but it pays the bills.

Starting from scratch

Starting from scratch with a basement conversion is often easier than fixing what already has been done. This type of work pushes the limits of common cosmetic improvements, but it qualifies because the major structural components are in place. You have a roof, floor, and walls with which to work. We could devote several chapters to basement conversions without covering the subject completely. Therefore, we will keep our discussion aimed at the basics. Let's start at the bottom and work our way up.

Floors

Basement floors typically are made of concrete. A poured floor provides a solid base to work with, and many types of floor covering can be applied directly to the concrete. Vinyl can be glued to this type of floor, and tile can be set on concrete with great effectiveness. Some types of carpets are glued to concrete, and others are stretched over it. Tackless strips can be held in place by fasteners that are shot into the concrete. Wood floors require you to build some type of wood substructure to attach them to (Fig. 20-1). This usually is done with pressure-treated lumber held in place by concrete fasteners, usually fired by powder-actuated nailing guns.

Moisture frequently is a problem in basements. Some basement floors flood annually, and many basements become musty. A dehumidifier will take care of the mustiness, but a more aggressive water-control system is needed for basements in which standing water appears. Installing slotted drain pipe around the perimeter of a basement isn't a big job in terms of technical skill, but it requires a lot of work. A channel has to be made for the piping, and this is normally done with a jackhammer. The pipe is laid, on a pitched grade, so that it drains into a sump. A sump pump then removes water from the sump and deposits it in an appropriate place away from the home. If this type of system is needed, make sure you install it before you begin your cosmetic work.

The type of flooring you install in a basement will depend partly on the type of room you are creating or reworking. Carpet usually works well in basements, and vinyl is sometimes a good choice. Tile does well in basement rooms, and wood can be okay. Because of possible moisture problems, I would hesitate to install expensive flooring in a basement before the space proved to be dry.

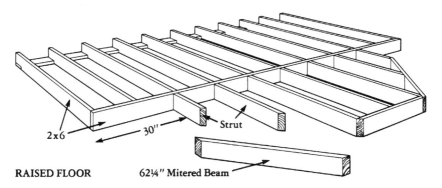

20-1 *A raised floor structure may be needed when a basement floor is remodeled.*

Ceilings

The ceilings in some basements are already finished off. This is often done with drywall, but ceiling tiles are common in basement ceilings. One determining factor is the number of obstacles, such as electrical wiring, plumbing, and heating equipment, that are suspended below the ceiling joists. It's common for workers to run pipes and wires below the ceiling joists in an unfinished basement. While this is fine when the basement is being left unfinished, the low-hanging devices cause a problem when a ceiling is wanted. This is why so many basements have dropped ceilings.

It's possible to move pipes and wires up into joist bays, but the work is time-consuming. The cost of relocating the items usually isn't justifiable. An exception is when the problem areas are small, such as in one corner of a basement. When this is the case, you can built a box around the items and make the rest of the ceiling flat. The boxed area can be disguised, with recessed lighting for instance. As long as you give the box a viable, cosmetic reason for existing, it will not be a problem.

Any type of ceiling material can be used in a basement. I prefer drywall. Some people like acoustic ceiling tiles. Wood can be used to finish off a ceiling. It's really just a matter of personal taste. Nine times out of ten, the ceiling of choice will be drywall.

Walls

The walls in basements can be bothersome. They often are made of cinder block or poured concrete., and neither material is easy to nail to. In some cases, a ledge exists where foundation walls end and stick-built walls begin. The ledge presents a problem because of its offset. And some basement walls stay damp much of the year.

Most contractors furr out basement walls before installing a finished wall covering. This, however, is not always the case. I've seen jobs where drywall and paneling have been glued directly to the foundation walls. One big problem with this approach is electrical wiring. Jobs work out better when the foundation walls are furred out.

If you are familiar with poured concrete walls, you know that they often contain humps and bumps. This is another good reason to furr them out. In the case of cinder block walls, you can use furring strips to build a wall out. The strips are enough to conceal wiring, and outlet and switch boxes can be recessed into the cinder block. Don't try this with solid concrete walls, however. When I figure a basement job, I always plan on furring the walls out with standard wall studs. This gives all of the trades room to work, and I don't have to worry about fitting plumbing, electrical, or heating equipment into a space that is too small to accommodate it.

When you must deal with a ledge, you have only two options. You can furr out the framed walls to be even with the new basement walls. Doing this creates a problem with windows, which are sunk so far back into the walls that they look strange. If you don't want this to be the case, you have to leave the ledge. I usually finish off the ledge with trim boards. Doing this creates a nice shelf for collectibles. In the case of a game room, the ledge can serve as a drink rail. Unfortunately, there just isn't much else that you can do with the ledge.

One thing that an alarming number of contractors fail to do is insulate basement walls. I've seen many jobs in which basements were furred out and never insulated. Since the basement walls are buried, I suppose contractors don't feel a need to insulate them. I disagree. A lot of cold can be transferred through concrete and cinder block. This may go unnoticed in an unfinished basement in which little time is spent. If you are creating living space in a basement, however, occupants of the house probably will notice the cold. It doesn't take much time or money to insulate basement walls, and it's an investment that will pay off for years to come.

Once you have a basement furred out, the wall coverings can be anything your customer likes. Since basements tend to be dark, it's best to work with materials that will not contribute to the darkness. If paneling is used, it should be a light color. Painted drywall always works well, and wallpaper is sometimes installed in basements.

Windows

Basement windows range from small, unimpressive peepholes to nice, full-size windows. When working with daylight and walk-out basements, you have normal window options. This is not the case with buried basements. Natural light is important, but you can't get much of it in a buried basement. What can you do with those little rectangles that are placed high on basement walls and considered windows? Not a lot, but there is one little trick that I've used a few times.

The small, rectangular windows found in most buried basements don't provide any substantial benefit in terms of natural light or of being able to gaze out at the great outdoors. Since the windows provide little practical purpose, why not make them decorative? Ah, but how can a squatty little window be made decorative? Use stained glass covers over the existing glass. The colors and designs used in stained glass make for a pretty window.

An office manager who worked for me many years ago made stained glass as a hobby. As a Christmas gift, she gave me a beautiful piece of stained glass that was very colorful. She had made the glass in a diamond shape, to fit the window in my foyer. All I had to do was slide the glass into the frame of my window and secure it, using tiny nails as stops. The new look make my foyer much nicer. After seeing what a difference the stained glass made in my foyer, I got to thinking. It seemed like the idea of using stained glass in some of my building and remodeling projects made sense.

When I remodeled a basement to include a finished family room, I immediately thought of the stained glass. I asked my manager if she would be interested in making a few pieces of glass on a custom basis. She was pleased to do it. After talking with my customers, we decided on a pattern with upland game birds in it. Colors were selected and a rough sketch was given to my manager, along with a template of the window sizes.

When the family room was finished, I installed the stained glass in the windows. It was attractive, and the customers loved it. As my customers had friends over, the windows got some public exposure. It wasn't long before people were asking me to put stained glass in their basement windows. I even got calls for cabinet doors with stained glass in them. My manager was happy, because she was making more than enough money from her hobby to allow her to do more of what she enjoyed without dipping into her regular paycheck for supplies. The use of stained glass generated a lot of attention, and I'm sure I got some jobs just because the people started talking about the glass and wound up calling my company for remodeling work.

Doors

Exterior doors in walk-out basements often are sliders. This is good from the point of view that sliding-glass doors allow a lot of light to enter a room. But many builders have installed cheap sliders in basements. The mentality seems to be something along the lines of, "It's just a basement, so there's no reason to install good doors." Many sliding-glass doors cause problems. In winter, some doors collect condensate to the point where frost and ice build up on the frames. Cheap doors usually don't slide very well. This can be frustrating for anyone, and it can be very bad for children or elderly people who lack the strength needed to force the sticking doors open.

When I've run into problems with sliding doors in basements, I've usually replaced them. Sometimes the replacements have been better sliding doors, but my most common replacement door has been a two-panel, swinging terrace door. Only one panel opens, but the unit resembles a pair of French doors in appearance. I normally use metal, insulated doors with plastic grids to give the impression of panels of glass. This type of unit is not cheap, but it is affordable. Energy efficiency is good, and almost anyone can open the doors with ease.

Basements can present contractors with plenty of challenges, but they also offer vast opportunities. While basements are not always considered good places to invest improvement dollars, they are a space that often is converted into living space (Fig. 20-2). Don't overlook the many possibilities offered by basement remodeling.

20-2 *An example of a successful basement conversion.*

21

Dining rooms

Dining rooms aren't as plentiful as they once were. More and more new houses are being built with eat-in kitchens, thus avoiding the cost of separate dining areas. Houses that have dining rooms vary in size and value. Some dining rooms are formal, while others are merely rooms designated to hold a table and chairs.

As a cosmetic contractor, there are two basic scenarios you are likely to encounter when working with dining rooms. You either will be refurbishing an existing formal dining room, or converting a informal dining room into a formal one. You could be asked to build an addition to make space for a dining room, but I will limit this discussion to cosmetic work in existing dining rooms.

In many ways, formal dining rooms can be compared to formal living rooms. Since this is the case, I won't go over, item by item, the information in Chap. 16. Refer to that chapter for basic information on formal living rooms and apply the information to dining rooms. There are, however, some differences between living rooms and dining rooms. Let's talk about them.

Chair rail

Chair rail should be installed in dining rooms, not only because the trim is attractive, but because it is functional. The rail keeps chairs from gouging walls, a common problem in dining rooms, especially small ones. When I think of chair rail, I tend to see it installed as a divider between two types of wall coverings, but that does not have to be the case. Chair rail can be installed on walls that are painted from floor to ceiling.

It's been my experience that customers sometimes balk at the idea of installing chair rail, usually for financial reasons. When you look at the cost of chair rail, it's easy to see why it might intimidate some people. Even though chair rail does have a high per-foot cost, most dining rooms don't require a lot of the material. Forty feet of chair rail will be more than enough for some dining rooms. If you're dealing with big houses, the cost can get uncomfortable. With this in mind, let's talk about room sizes.

Room size

The size of a dining room has some effect on the cosmetic conversions done in it. Some modern dining rooms are very small (Fig. 21-1). Others are an average size. In the case of old houses, dining rooms can be huge. When I think of a big dining room, I see the one that I ate in so often at my grandmother's house. The visual image of that multipurpose dining room is clear to me. It was so big that it would cost a small fortune to do modern cosmetic improvements on it.

My grandmother's dining room had a large chandelier in the center. A very big table was placed under the fixture. One corner of the dining room was home to a full-size piano. China cabinets and a pie safe took up one of the walls. A sewing machine and daybed were placed along one of the outside walls. An old oil-burning space heater sat at the common wall between the dining room and living room. During December, the dining room always had a large Christmas tree standing against one wall.

I never measured the dining room in my grandmother's house, but it was larger than the living room in my parents' home. The ceiling was tall and covered with decorative tin. Wallpaper covered every wall from floor to ceiling. When Thanksgiving rolled around, the room would handle more dinner guests than some small cafes. My grandmother's dining room was larger than normal, but I remember the dining rooms in other houses of that same age that also were very big.

As a remodeler, I've never tackled a dining room of mammoth dimensions, but they're out there. I would say the average size of the dining rooms I've remodeled would be in the neighborhood of 125 square feet. A room of that size is puny compared to the dining rooms in some old farmhouses. Paying to install fancy trim in some old dining rooms could cost more than putting a new floor in a modern dining room. Since the cost of materials is relative to their quantity, you must consider this when planning a cosmetic conversion. Something like chair rail, that might be too expensive in a large room, can be very affordable in small rooms.

As difficult as large rooms can be to work with, small rooms are, in ways, harder to deal with. Assume that you are called to rework a small dining room. For the sake of discussion, let's say the room is about 8 × 10. The customer wants you to freshen up the room with

21-1 *This is a modern dining area that adjoins a kitchen.* Congoleum Corp.

new paint and carpeting. During your talk with the customer, you find she has a collection of china that has been handed down from family to family. Since the dining room is so small, the customer hasn't been able to put the china on display. Instead, it collects dust in a kitchen cabinet, over the refrigerator. Think for a minute. Is there anything you could suggest that would get the china on display in the new dining room?

Since the customer feels the room is too small to accommodate a china cabinet, your options are limited. But there is a solution to the problem. Suggest building corner display racks. By cutting triangular shapes of wood and installing them in one or two corners, you can build a display area that will not take up much usable space. The triangular holders can run from floor to ceiling. With some turned dowels as railing, the display would be very attractive. Cleats could be installed to separate and secure plates. This kind of creativity goes a long way toward making a small room more enjoyable.

Floors

Dining room floors normally are carpet or wood (Fig. 21-2). Hardwood floors are nice, but also are expensive. Most people agree that carpeting is an adequate substitute for hardwood flooring in a dining room. Yet I see more dining rooms with hardwood flooring than any other room in homes. I think this is due both to tradition and to the fact that most dining rooms are not so large that the installation of hardwood flooring will break the bank.

I can't remember ever being asked to remove carpeting and install wood flooring. As a builder, I've gotten many requests for wood floors in dining rooms, but remodeling hasn't produced the same results. People tend to stick with the type of flooring already installed. Some folks even opt for vinyl flooring, and it doesn't look bad (Figs. 21-3 and 21-4).

I can't see justification for covering up hardwood floors with carpeting, but I can see removing carpeting and replacing it with wood (Fig. 21-5). If I were converting an informal dining room to a formal one, I would consider this option. My feelings on this subject probably differ from those of my customers. Even though I have suggested the replacement of carpet with hardwood several times, no one has taken me up on it for a dining room.

Dining rooms get very little foot traffic compared to other parts of a home. Many families use their dining rooms infrequently, taking meals in their eat-in kitchens instead. Even if a dining room is used daily, the foot traffic is minimal compared to that on floors in kitchens, halls, bathrooms, living rooms, or family rooms. The low amount of traffic increases carpeting options. You can get by with a carpet of lower quality when traffic on it will be light. Most customers, however, go for an upper-end carpeting for dining rooms. Some of the logic customers apply to dining rooms doesn't make sense to me—but hey, the customer is always right.

I've found that many customers prefer light, solid colors for carpets in their dining rooms. I've done a number of rooms in stark white colors. Tan and beige have been popular with my customers.

Dining rooms

21-2 *This dining room has a wood-grain vinyl floor.* Azrock Floor Products

Royal blue has been used in more than one dining room that I've redone, and so have maroon colors. The darker colors tend to make a room look smaller than lighter colors do. Another consideration with color is the transition point between the dining room and other living space. If a deep, rich color is installed in the dining room and butts against a neutral color in an adjoining living room, the contrast will be extreme. Try to avoid this.

Tile, such as quarry tile, is gaining some popularity in dining rooms (Fig. 21-6). Some customers might prefer a tile floor that is a

21-3 *Vinyl flooring in a dining room can look nice.* Mannington

bit more colorful (Fig. 21-7). In any event, don't rule out tile as a possible flooring choice for dining areas.

Ceilings

Ceilings in most dining rooms are done with drywall. They may be painted or textured. Some sort of fancy ceiling is used occasionally, but drywall is, by far, the most common type of ceiling material. There are certainly other options available for dining rooms, but I can

Dining rooms

21-4 *Don't rule out vinyl flooring when a dining room is remodeled.* Mannington

see no reason, under normal circumstances, to stray from drywall. Exceptions to this might be made when working in older homes or houses with elaborate finishes throughout. You might even find occasion to install skylights in a vaulted ceiling (Fig. 21-8).

Walls

I don't think you can go wrong with wainscotting, chair rail, and wallpaper when dressing up a dining room (Fig. 21-9). Plain paint is fine,

21-5 *Wood flooring looks good in both traditional and contemporary dining rooms.* Pella

but wallpaper makes a more formal statement. If your customer doesn't like or can't afford wainscotting, consider painting the lower sections of walls and papering the upper sections. When you have customers who want to keep their jobs simple and inexpensive, go with a good grade of paint in a neutral color.

Doors

Doors may not be a consideration in some dining rooms, but adding a door to separate a dining room from other living space has advantages. One reason for adding a door is privacy. Security from family pets that might raid the dinner table is another viable reason for adding a door. If your customer has the space and the finances for it, a French door is always a nice touch for a dining room. Doors aren't essential, but they can add considerably to the overall effect of a dining room, especially a formal one.

Have you ever added exterior doors to a dining room? I have, and they can work out nicely. The addition of an exterior door — terrace doors are my favorite — and an outside deck adds use and meaning to

21-6 *Quarry tile can be used to warm a dining area.* The Tileworks

a dining room. Light can flood in through the door, and the deck gives people a place to roam before and after a meal. When the weather is right, cooking out on the deck is fun, and the dining room is close by. Opening up an outside wall with doors and windows that lead your eye to a deck will make a dining room feel larger. In good weather, the deck serves as a meeting place where guests can gather while waiting for the dining room table to be set. The smooth flow between a dining room and deck makes a small room more inviting in good weather.

21-7 *Tile patterns can vary greatly, and their selection does play a vital role in the appearance of a room.* Summitville Tiles, Lis King

Windows

Windows are important to most rooms, and dining rooms are no exception. Most dining rooms have at least one outside wall where windows can be placed. Many dining rooms have two outside walls, and this gives a builder or remodeler plenty of wall space with which to work when installing windows. It is normal for a dining room to have the same type of windows as has been used in most of the other parts of a home. There's nothing wrong with this, but you and your customer should consider making some alterations when contemplating cosmetic improvements. The types of windows you might use will be determined by the style and size of the dining room.

Small dining rooms can't accommodate a lot of glass. When typical dining room furniture is in place, most of the wall space in a small room will be used up. There's not much point in having windows that are hidden behind large pieces of furniture. When you are working with a room that is large enough to allow more than normal ventilation, you have lots of options. And there is one type of window arrangement that can actually add space to a small dining room.

Assuming that you are dealing with a room in which space is not a problem, you might remove standard windows and replace them

21-8 *Skylights are a nice touch in modern dining areas.* Velux

with very tall windows. Arched transoms could be added. Walking into a dining room through French doors and seeing a wall of tall windows is a powerful feeling. You can almost imagine that you are dining anywhere. Of course, a home's surroundings have something to do with how much glass would work. If the view will be of a neighbor's house only a few feet away, it might not make sense to invest in glass. However, if the view will be of woods, meadows, mountains, or the seashore, the more glass the better.

Chapter Twenty-one

21-9 *Chair rail is a wonderful addition to any formal dining room.* Georgia-Pacific

I've worked in crime-infested cities and beautiful country settings. My jobs have been in townhouses, zero-clearance zoning areas, and on vast tracts of farmland. Each setting requires its own type of consideration. Since I live in the country and prefer rural surroundings, I tend to think in terms of a nature-lover's setting. But, I'm aware that my ideal setting is not typical. Most contractors probably deal with houses that are close together and where privacy comes in the form of fences or blinds instead of from a natural shield of evergreens and birch trees.

The fact that a house is in a city setting doesn't mean that it shouldn't have windows in the dining room. Many urban homes have nice lots and enough distance between houses to warrant large windows. You will have to consider the location of a job when offering ideas, but at least consider the possibility of installing additional windows in dining rooms.

I mentioned a type of window arrangement that could add space to a small dining room. The window type to which I was referring is a *walk-out bay window*. This type of setup is not cheap to build. A foundation usually must be installed to support the walls of the walk-out area. You could say that this is a small addition to a home, and it is. By installing a walk-out bay, you can add a little square footage to a room, making the space appear much larger.

When you add a walk-out bay window, the additional floor space can be used for a number of things. It creates a place to park a serving table. Chairs can be placed in the window area for before-and-after dinner chats. A potted plant can grow in the window area. Hanging baskets of plants and flowers can be suspended from the ceiling of the window. The construction of a window seat can provide both seating and storage. There are so many possibilities for walk-out windows that they should be given consideration.

Lighting

Traditional lighting in a dining room calls for a chandelier to be mounted in the center of the room. This one light fixture often provides adequate light. If more light is needed, consider having wall-mounted lights installed. Track lighting can be used in modern, informal dining rooms, but carriage lamps might be more appropriate. Adding a dimmer switch to an existing chandelier can make the room more romantic for those special meals when bright lights are not needed.

Many people use their dining rooms as a place to exhibit family portraits. I've seen dining rooms in which three out of four walls were covered with pictures. When this is the case, you might want to consider building a small bulkhead over the photos so indirect lighting can be installed to illuminate the pictures. Little suggestions, like this, can mean a lot to customers, even if they don't choose to include the option in your work.

Identifying your path

Before you begin making cosmetic conversions, identify your path. Is the room meant to be formal? Do your customers want a contemporary look? Should the room be kept plain and simple? Are the wishes of your customer reasonable considering the size of the room? Some contractors don't ask questions. They listen to what customers think they want and agree to do the work. I feel it's best to make sure customers understand what their finished product will look like. After all, cosmetic conversions are all about appearance.

It's always important to make sure that you and your customer understand each other. Dining rooms offer a lot of room for mixtures that might not look good. Before you start reworking a dining room, check every detail. For example, track lighting will look fine in a contemporary room, but it would look out of place in many formal dining rooms. If you go over your plans carefully, you can check with customers on any ideas that you feel might be inappropriate. Give customers what they want, but make sure they understand what they are getting. This will make your life easier and your customers happier.

22

Laundry rooms

Laundry rooms are probably one of the last rooms that would come to mind if you were asked to conjure up images of cosmetic conversions. Granted, laundry rooms aren't usually seen by guests, but homeowners have a right to be happy with their homes, and a good deal of time can be spent in a laundry room. While I would expect a laundry area to be low on the list of priorities for most homeowners, I know for a fact that people do ask contractors to perform cosmetic conversions on laundry rooms.

I've seen some very ugly laundry rooms. Most laundry areas are plain and unimpressive. This usually isn't a big problem, but some people use their laundry rooms as more than just a place to store dirty clothes and cleaning equipment. In fact, I've had jobs where my goal was to enlarge a laundry area to include space for a sewing machine or some other type of equipment. One homeowner had me expand a laundry room to include a photography darkroom. Until your phone rings with a request, you never know what to expect next.

Types of laundry rooms

There are many types of laundry rooms, and all have a common function: getting clothes clean. But the size, shape, and location of a laundry room can vary greatly. Many houses are built with what I call closet laundry areas, with the washer and dryer are placed in a small closet and concealed with bifold doors. This type of laundry room doesn't offer a cosmetic contractor many options. You can paint the walls, add a shelf or two, and replace the door or flooring, but that's about it. Since closet laundries are common, I'll describe ways to make them more useful, but it won't be a long discussion.

Houses with basements often have laundry areas in some part of the basement. The space dedicated to laundry work usually is left unfinished. A good aspect of basement laundries is that they often provide plenty of space with which to work, and that's a big advantage.

When a customer wants to have work done to beautify a basement laundry, a cosmetic contractor can generate a lot of work out of the job.

It's not unusual to find homes that have laundry areas in a garage. The room usually shares a common wall with interior living space. Access to a garage laundry might be possible from the garage only, from the interior of the home only, or from both places. There are some advantages to this type of setup. If people come home dirty or wet, they can come in through the garage and dump the wet or dirty clothing in the laundry area. This keeps the mess out of the main part of the house.

Garage laundries usually can be expanded easily. Depending on whether a garage laundry is accessed directly from a home interior, the finish work in the laundry might be normal or crude. If the only access is from the garage, it's not uncommon for the walls, ceiling, and floor to be rough. As a cosmetic contractor, you might be asked to improve the finish work and add a door to the main part of the home. I haven't seen a lot of new houses with garage laundries, but I know of many existing homes that have them.

The last type of laundry room to be discussed is the in-house laundry. These laundry rooms occupy space in the main living area and are full-size rooms. The room size might not be as large as a bedroom, but it will be large enough to move about in freely. In-house laundry rooms are somewhat rare in modern homes. Builders tend to create tiny laundry areas in order to make other rooms larger. I think that can be a mistake.

As a home builder, I've built houses with all types of laundry areas. In fact, I've probably built more homes with closet laundries than with any other type. But I've built a lot of in-house laundry rooms, too. There has been, and still is, a demand for a comfortable room in which to do laundry work. As a matter of fact, I had a customer who contracted me to move her laundry from her basement to the main living area. She sacrificed part of a bedroom to get an upstairs laundry room. So you see, some people will go to great lengths to get the type of laundry room they want.

Closet laundries

Closet laundries don't offer much in the way of cosmetic conversions. When the doors to this type of laundry are open, there isn't a lot to see. A little flooring is exposed, and there are some wall and ceiling areas. Paint usually is the only logical cosmetic approach to the walls and ceiling in this type of space. New flooring might be needed, but it need not be fancy or expensive.

When working with closet laundries, you are likely to be working with add-ons or built-ins. A customer might want you to add a storage shelf or two. Maybe you will build a wall-mounted rack to hold detergent, bleach, and other cleaning materials. Someone might want you to install a ceiling light. All in all, there's just not much that can be done with a closet laundry.

Basement laundries

Basement laundries can run the gamut from crude to comfortable. Many basement laundries consist of nothing more than a plumbing hookup for a washing machine and a vent connection for a clothes dryer. Electrical devices include outlets and maybe a pull-chain light. In many cases, there is no covering on the concrete floor, no finished ceiling, and the walls are either concrete or block. You will find a laundry tub near the washing machine in some instances, but that's generally the extent of a basement laundry. Even when a basement is partially finished into living space, the laundry is not usually a part of the finished space. If this is the type of laundry room you encounter, there's plenty you can do to improve the work space.

I've talked about methods for finishing basements, so that material won't be rehashed here. Refer to Chap. 20 if you need a refresher on materials and methods for basement conversions. Let's focus on the options pertaining specifically to laundry rooms.

Assume that you have been called into a job that is similar to the laundry area I've just described. The homeowners want to make the area a more pleasing place. At the same time, they would like to make the area usable for more than just washing and drying clothes. What can you offer them?

The first step will be building walls for the room. Finish wall coverings will probably be drywall and paint. The ceiling, if there are no low-hanging obstacles, also will be done with drywall and paint. A dropped ceiling might be needed, and this can be done with ceiling tiles or false framing and drywall. Sheet vinyl makes sense for the finish flooring. The door to the room can be any standard interior door. This will give you an enclosed space, but what will you add to the area to make it more comfortable and more efficient?

There are many suggestions that you might make for built-ins and add-ons. If the room doesn't have a laundry tub, consider adding one. The plumbing will not be difficult, and a laundry tub is handy. It can be used for everything from soaking stubborn stains to washing the family pets. A built-in, drop-down ironing board would be a wise consideration (Fig. 22-1). It will save space and be convenient. What else can you do?

22-1 *This is a well-appointed laundry room.* WilsonArt

Shelves, or even a storage closet, could be added to the space. Every laundry area needs some storage, and the more the better. A built-in, drop-down sorting and folding table could be helpful to a person taking care of laundry duties. Like the ironing board, a built-in table will be out of the way until it's needed, and it can be put to use quickly and easily.

A sewing area might be appropriate for the laundry room. Many people sew as a hobby and others sew out of necessity. In either case, a designated sewing area can be a real boost for some homeowners. It doesn't take much to create this type of space, but the convenience of being able to leave a sewing machine set up, ready for use, is a plus for people who sew.

How many people have you heard complain about the storage of off-season garments? I've talked to countless people who complain about not having enough closet space. Much of their problem stemmed from having clothes for all seasons stored in one place. Since basement laundries tend to have plenty of room with which to work, consider building a storage area for off-season clothing. As the seasons change, clothes can be cleaned and stored without leaving the laundry room.

Clothes drying on a line is something you don't see much of anymore. It used to be common for people to take a basket of clothes outside and pin them to a string or rope. Many times the clothesline conversations that went on between neighbors were an important part of the day. With the use of electric clothes dryers and a lot of couples working at full-time, out-of-home jobs, the use of clotheslines has diminished. Some garments, however, dry better when hung on a line. With this in mind, you might consider installing an indoor clothesline in a basement laundry. Having this line for special-care garments could be useful.

Garage laundries

Garage laundries might be similar in appearance to basement laundries. You might have a concrete floor to cover, and the room might not have a finished ceiling. The walls might be bare studs. When this is the case, you are starting from scratch and have a full array of cosmetic options with which to play. On the other hand, you might find garage laundries that are finished in much the same way as the rest of the house. You also could find a garage laundry at some stage in between, for instance with a vinyl floor, and walls and ceiling that have been drywalled and taped, but not finished or painted.

If a customer wants to expand a garage laundry and is willing to give up some garage space, the expansion is usually easy. It's often just a matter of moving nonbearing partitions around. In some cases, floor heights might differ, complicating the expansion, but not to a point that extreme cost is incurred.

All of the ideas given for basement laundries can be applied to garage laundries. In addition, you might consider adding an additional door to the laundry room. If it has a door only between it and the finished living space, think about installing a door between the laundry and the garage. Having a garage laundry accessible from either the garage or house is a convenience that most homeowners won't refuse.

In-house laundries

In-house laundries can be difficult to expand, because any expansion probably will affect other finished living space. Unless an in-house laundry room shares a common wall with a garage, expansion is usually out of the question. This doesn't mean that there is nothing a cosmetic contractor can do with an in-house laundry room. To illustrate, let me give you an example of a job my company did several months ago.

I was called to a home with an in-house laundry to give an estimate on cosmetic work. The room was not large, but neither was it tiny. No one had ever vented the dryer to outside air, so moisture had taken its toll on the paint in the room. An offset in the room appeared to have been made for a laundry tub, based on the dimensions of the space, but no tub was present. After looking the job over, I made several suggestions.

One of the first changes we made in the laundry room was the addition of a laundry tub. Then we built large shelves over the plumbing fixture. Shelves were installed above the appliances in an L shape to provide even more storage area. A pull-out bin with several drawers was installed beside the washing machine in a little dead space that existed. Since the vinyl flooring was in good shape, we didn't do anything with it. Next, we prepared and painted the walls, ceiling, and trim. Coat hooks and other wall accessories were added to the walls to hold the coats, hats, and backpacks of children who used the room as a drop-off point when they came home from school. A boot rack was installed under the laundry tub. When we were done, the room was no larger, but it had more space. The hooks, racks, and shelves did a lot to clean up the floor space. While still cramped by my standards, the laundry room was much better than it had been when we started our work.

Cosmetic work in laundry rooms is rarely glamorous. A remodeled laundry room isn't usually the type of job contractors brag about or show pictures of. But reworking a laundry room can be difficult, and the work deserves some credit. I think one reason that so many laundry areas are ignored is that they aren't public-orientated spaces. Few homeowners take guests on a tour of their laundry facilities. It should be recognized, however, that a laundry room should be as comfortable to work in as is a kitchen. Kitchens get all the fanfare while laundries are ignored. This shouldn't be the case, but it often is. The next time you get a call to remodel a laundry room, take the job seriously and put some effort into creating a nice work area for your customer.

23

Sun rooms

Sun rooms are among my favorite types of rooms to remodel. The last two houses I've built for myself have had sun rooms, and I have enjoyed them so much that I wonder why I settled for decks in my previous houses. Sun rooms are great for my lifestyle. I love the outdoors and live in the woods; having a sun room allows me to breakfast with the forest at any time of year. If you're thinking that I'm a little weird, you're probably right, but many homeowners do share my appreciation of sun rooms.

Some sun rooms are plain and simple, while others are fancy. Many sun rooms started out as decks or screened porches. Homeowners sometimes tackle this type of project on their own, creating working conditions that some contractors would rather avoid. I've seen many sun rooms, and there is usually room for improvement with most of them.

Some sun rooms started out as open decks, with the decking serving as a subfloor. Others once were covered porches with concrete floors. Many sun rooms have been built on pier foundations. The flooring in an existing sun room might be indoor-outdoor carpet laid over deck boards, expensive tile laid on a concrete base, or luxurious carpet laid on a plywood subfloor. There can be a lot of difference in the quality of materials used to build sun rooms, and that applies to more than just the flooring.

Walls in most sun rooms feature a lot of glass; some are built almost entirely of glass. Curved glass panels can be used to make the walls with a continuation that creates a roof. Sliding glass doors are often used as wall panels for sun rooms. For people who want electrical outlets in walls and some privacy, half walls can be built up to a drink rail; glass takes over from that point. Walls that are not glass can be painted, paneled, or wallpapered.

Sun rooms often have vaulted ceilings, with a ceiling fan hanging from the center. Exposed beams are a common decoration in sun rooms. The ceilings themselves are sometimes made with T&G planking, but drywall ceilings also are common. Rooms that don't have glass roofs might have skylights. When you consider all of the elements that go into making up a sun room, it's easy to understand why these rooms appeal to cosmetic contractors. There is a lot that can be done with a sun room.

People often use their sun rooms as entertainment areas because they offer an informal setting where guests and family members can relax. Because visitors often are invited into sun rooms, homeowners tend to want the rooms to look good. This is advantageous for cosmetic contractors. People who are concerned about the appearance of their sun rooms are likely to keep the rooms up to date and in style.

Let me share the experience of one of my sun room jobs, a sun room that had been made as a conversion from a screened porch. Licensed contractors had done the conversion work, but the finished product left a lot to be desired. I can only assume that the homeowners who authorized the work had very little money to spend. When the house was sold to the people who hired me, they decided to upgrade the room.

The original screened porch had been built with a pressure-treated floor. The material used was square-cut, not nosed. Screen had been installed under the flooring, and was stapled to the floor joists to prevent insects from entering from below. When the porch was transformed into a sun room, the only modification made to the floor was the addition of a layer of green, turf-type carpeting. Cracks between the floorboards were evident through the carpet (I use the term loosely). Some of the boards had shifted to the point that they were a tripping hazard, even with the carpet in place.

During the original conversion, drywall had been nailed to the lower half of walls, creating the sun room. This part of the job was in pretty good shape. Rough-cut boards served as finish trim along the baseboard. Even these didn't look too bad, though. It was the glass sections that made me shake my head.

The windows used to create the glass walls were cheap aluminum sliders, the type that accumulate condensate and freeze up during winter. Not only were the sliders old and cheap, their very design defeated some of their purpose. Only half of the window could be opened, so air flow was restricted. Oddly enough, the contractors who built the room basically stacked one window on top of another to create the wall of glass. The windows on the upper tier couldn't be reached by the homeowners, or me, without the use of a step stool.

The ceiling in this sun room was made with a thin wood paneling that was coming apart due to moisture. It was nailed right to the ceiling joists of a shed roof. There was neither skylight nor ceiling lights.

The porch had been built onto the side of the house. When the conversion was done, no attempt had been made to finish a wall where the addition attached to the home. Consequently, one wall of the sun room was made up of exterior siding on the home. Electrical outlets were few and far between, and the room was sitting on a pier foundation. There was a substantial stepdown between the sliding-glass door from the home and the floor level of the sun room. I had my work cut out for me on this one.

Since we are concentrating on interior rooms, I won't go into a lot of detail about the improvements we made to the exterior of the sun room. Basically, I enclosed the pier foundation with pressure-treated lattice and replaced the 4×8 sheets of siding with clapboard siding that matched the rest of the house. We replaced the exterior door and steps, too.

When I assessed the floor of the existing room, I was tempted to cover the existing floorboards with plywood and let that be the subfloor. A closer inspection revealed that it would be impossible to make a level floor that would be stable if I left the old floor in place. Based on this, I had my crews strip out the old carpet and floorboards. We were about to install plywood subflooring when I thought about the large stepdown into the room. I consulted with the property owners and gave them the option of keeping the stepdown or having us raise the floor level in the sun room. They liked the idea of having the floor at the same level as the floor in the house. To accomplish this, we built a new floor structure on the existing floor joists. It only took a few hours and the materials needed didn't amount to much. The new joists were covered with plywood for subflooring.

The lower halves of the walls in the sun room were in good shape, but they didn't have adequate electrical wiring. This was something we had to bring up to current code requirements. Rather than snake and patch, we stripped the existing drywall and exposed the wall cavities. This made doing the electrical work easier, and getting rid of the drywall was no big deal.

I had my crews frame a wall that would hide the siding of the house that was at one time a finished wall in the sun room. All windows and upper framing were removed and walls reframed to accept new casement windows. Since casement windows open out, both panels can be open simultaneously, providing twice the ventilation of a sliding window. Instead of stacking two windows on top of each other, we used casement windows that were large enough to fill the entire upper wall area with just one window unit in each location.

When we removed the drywall from the lower parts of the walls, we discovered that the wall cavities had never been insulated. We took care of this problem before we closed the walls with new drywall.

The old ceiling was removed completely. Three skylights were installed in the roof, and drywall was used as a finished ceiling material on the vaulted ceiling. The ceiling was completed with a textured finish, and a ceiling fan with a light kit was installed to provide light, air circulation, and aesthetic benefits.

Once the casement windows were installed, finished trim wood was used to dress up the little pieces of wall space between the windows. All of the trim was stained in a dark color, while the ceiling and lower walls were painted in an eggshell color. My plumber installed a sink for a wet bar in one corner of the room, and an independent GFCI electrical circuit was installed in another corner, to accommodate a spa.

The floor of the sun room was finished with large tiles that were white with blue designs. Colonial trim used for baseboards and door casing was stained in a dark, walnut color. Wood cover plates were used for electrical switches and outlets. The old sliding glass door that used to join the sun room with the main house was replaced with a terrace door. Since this door was metal, it was painted rather than stained. Another metal insulated door was installed as an outside exit for the sun room. This door had a full glass panel with plastic insert grids to simulate panes.

My customers spent a lot of money to get their old sun room brought up to date. Their investment was, in my mind, worth every penny spent. The new sun room didn't even resemble the old one. Not all homeowners are willing or able to make such major improvements in their sun rooms, but it certainly is nice to see a desirable room develop before your eyes. What my company created was more than just an updated room. We gave out customers a customized retreat from the stress of a busy life.

A room of many purposes

Sun rooms can have many purposes. I did a sun room conversion for a woman who wanted to use the room for growing flowers. She wanted more of a greenhouse than a sun room. Many of the basic building components were similar, but due to the specialized use, customized alterations were needed. For example, we had to provide climate control and ventilation. The glass had to be capable of being shaded easily, and a laundry tub was installed to provide water and a potting area.

Since watering plants can get messy, a slip-resistant tile floor was used to replace an existing carpet. The lower wall sections were drywall, and we put ceramic tile over the drywall to help make the room water-resistant. Instead of having glass walls and a conventional roof, as the existing room did, we removed the roof and replaced it with curved glass panels that served both as upper wall sections and roof. The glass panels were fitted with custom blinds. The result was a very nice sun room that doubled as a greenhouse. I remember another job where the homeowner wanted sun to flood the room, but didn't want neighbors to look in on her while she was sunbathing. For this job, we used glass panels for roofing and a short upper wall section. The rest of the room was converted to a conventional room, with casement windows that could be blocked with blinds. This arrangement allowed the woman to block off the sight lines by closing the blinds on the casement windows, while still receiving a lot of sun, at certain times of the day, through the roof and high wall sections.

Many of the sun rooms that I've remodeled have had sliding glass doors that served as walls. What I don't like about this is that only half of the wall section can be opened. My preference is to use casement windows, but this is a more expensive option. Sliding glass doors are a fast, affordable way to make a sun room. One job that comes to mind is a covered porch that we converted to a sun room. The porch had a concrete floor and a roof structure, complete with ceiling. About all we had to do was add a little framing and a lot of sliding glass doors. The conversion was inexpensive, in relative terms, and it suited the customers.

Before you start doing cosmetic conversions on a sun room, you should talk with your customers to determine the intended use. Knowing what the customers want to do with the rooms can be a big help in sorting through the options available. Some people use their sun rooms as a place to enjoy a quiet breakfast. Many homeowners entertain guests in their sun rooms. I've done a lot of sun rooms in which the goal was to install a spa and have a wonderful place to kick back and relax. I've done a few conversions that were undertaken to create a cheerful playroom for children. Each use will have different material requirements.

Floors

Floors in sun rooms range from indoor-outdoor carpet to fancy tile. I've even seen hardwood flooring in some sun rooms, but this is unusual. Carpeting is popular, and I would say it is the top flooring choice for sun rooms. Personally, I prefer tile. If a sun room is being used during cold weather, tile floors can help to retain heat that is gained through the glass. Before you settle on a flooring, you and your customer have to determine how the room is to be used. Carpeting often is the most logical choice.

Walls

The walls of sun rooms sometimes are all glass. This requires the use of floor-mounted electrical outlets and doesn't allow for any wall insulation, beyond that of the glass. Full-glass sun rooms are expensive, and can pose some risks or problems. Parents who have young children are frequently concerned that their kids will break the glass. I feel that it would take some extreme conditions for this to happen, but it is a possibility. Except for jobs in which I've used sliding glass doors as walls, I would say that the majority of my sun rooms have had solid walls half way up, and casement windows or fixed glass panels in the upper wall sections.

Fixed glass is the least expensive material to use for glass walls. One problem with this approach, however, is that the glass cannot be opened. Consequently, a sun room can get very hot. If the room is air-conditioned, this might not be a problem, but I prefer glass that can be opened to allow ventilation. Still, to keep costs down, I've done a number of conversions with at least some fixed glass panels.

The lower wall sections in sun rooms can range from tile to drywall to wood. One of my favorite materials is T&G planking. It can be installed in a number of patterns. Herringbone is one of my favorites. Even when the planking is run horizontally, it looks very good. Drywall is, of course, the least expensive option. As much as I like tile, I prefer not to use it as a wall finish in sun rooms. Given a good budget to work with, I opt for wood every time. Since sun rooms are bright to begin with, you don't have to be concerned about the wood, whether it be planking or paneling, being too dark.

Ceilings

Ceilings in sun rooms can be flat or vaulted. I prefer vaulted ceilings with operable skylights. Whenever I can, I install ceiling fans with light kits in sun rooms. The fans are both attractive and functional. Painted or textured drywall is the most common ceiling material in sun rooms.

When working with a vaulted ceiling, I often suggest the use of exposed beams. These beams are usually false beams, made of either trim boards or some prefab beam material. I also like to install T&G planking as a ceiling in sun rooms. Many customers prefer narrow planks; others like wide ones. To me, a plank ceiling with exposed beams, skylights, and ceiling fans is hard to beat. Unfortunately, this type of upgrade is expensive, so most ceilings wind up being painted drywall.

Lighting

Lighting in sun rooms can be accomplished with track lighting, ceiling fans with light kits, regular overhead lights, recessed lights, or other light fixtures. I prefer ceiling fans with light kits and track lighting. When prefab glass panels are used as walls and roofs, your lighting options are limited. You might have only one wall on which to mount traditional lighting. Keep this in mind when you are planning a solarium-type sun room. Artificial light usually isn't needed during daylight hours in a sun room, but don't forget that these rooms are frequently used after the sun goes down.

Plumbing

Plumbing is not a common appurtenance in sun rooms. However, I've remodeled sun rooms to include wet bars and plumbing for whirlpool tubs. Spas tend to win out over whirlpool tubs. Because spas are filled and drained with common garden hoses, fixed plumbing is not required. I would say that wet bars and spas are the most common plumbing fixtures installed in sun rooms.

Sun rooms are exciting areas to remodel. If a home doesn't have a sun room, you might be able to talk the homeowners into adding one. Don't overlook the possibilities of converting screened porches and roofed porches into sun rooms. There's a lot of money to be made with sun rooms, so they deserve a lot of attention.

24

Home offices

Home offices are popping up all over. Statistics indicate that more people are working from home more often than once was the case. Some are doing so on a part-time basis, but a lot of people are making the move from downtown offices to in-home offices. These offices are tucked into various areas of their homes. You might find them in garages, attics, basements, bedrooms, and even dining rooms. With the growing trend of companies trying to increase productivity and reduce expenses, home offices probably will continue to increase in popularity.

How do home offices affect you as a cosmetic contractor? You can benefit in two ways from the boom in home offices. If homeowners want to improve decor of their existing offices, you can help them. In cases in which homeowners want to create offices in their homes, you can cash in on the conversions. Either way, you win.

Potential office locations

Let's talk about potential office locations. If you get a call to create a home office, you must be prepared to offer options to your customers. In some houses, it can be very difficult to find extra space. Others, however, offer an abundance of opportunities for expansion. Where you create a home office for a customer will depend on many factors. Before getting into physical locations, let's discuss the factors that may lead you in one direction or another.

Before you can do a customer justice in offering office locations, you need to determine the office criteria. For example, will the office be receiving deliveries? What types of deliveries can be expected? Will the public be coming to the office? Is there a need for a separate entrance to the office? How large does the office have to be? Will employees be using the office? What will the equipment needs of the room be? These types of questions have to be answered before you can designate an ideal office space. Ask your customers what their needs are. Spend enough time discussing what will be wanted and needed to help determine ideal locations.

Once you have a checklist of criteria, you can begin to look at room locations. You might find space in an attic. Garages and walk-out basements can provide easy office space. Even daylight basements can be used for office arrangements. Other common office locations include unused bedrooms, dining rooms, and corners of other rooms. These locations are, for obvious reasons, more limited in scope than other areas of a home.

Access

Access to a home office can be an important aspect of site selection. Houses with walk-out basements can be converted so that office visitors enter the office area without going through the home's living space. Attic offices are more of a challenge. You can build stairs up the side of a house to get to an attic office. Depending on the layout of a house, reasonably direct access to an attic might be possible through a foyer. Unfortunately, it's difficult to get good access to an attic office. The addition of exterior steps usually detracts from a home's appearance, and they can become treacherous when covered with snow or ice.

Attic space works very well as a home office when heavy public traffic isn't likely. For home-based workers who don't have much in-office contact with customers and clients, attic rooms are fine. But take your business as an example. You probably make most of your customer contacts, other than phone calls, outside of your office. If this is the case, an attic office might serve you well. If, however, people come to your office to look at flooring samples and swatches of wallpaper or to discuss remodeling and building designs, an attic space probably would not be appropriate. This is the type of question you must consider when helping customers find suitable in-home office locations.

Basements with independent access can be used successfully as offices for all types of business needs (Fig. 24-1). You wouldn't want customers entering through a bulkhead door, but as long as access to a basement office can be accomplished with a full-size exterior door, the basement is a prime candidate for cheap office space. Where else could an office be fitted into a house?

Garages provide two types of office options. The first, and probably the most common, is to install an office in the attic space of a garage. This can be very effective. Access can be gained from outside stairs, which still have the drawbacks of poor appearance and questionable safety, or from interior stairs that take off from a foyer built into the garage.

24-1 *A comfortable home office in a basement.* _{Manhattan Cabinetry, Lis King}

It might not be logical to shoot for an attic office in a garage. The roof pitch and the roof structure might not be conducive to easy office installation. In that case, consider building the office space right in the garage proper. Your customer will have to forfeit parking space, but a garage can be converted into an ideal office environment.

Attics, basements, and garages are the best choices for expanding into home office space, but are not the only options. Some house designs lend themselves to office conversions very nicely. For example, you might be working with a house that has a bedroom or a dining room right off the foyer or entrance hall. Such an arrangement offers an instant office area with which to work. On the other hand, if clients must walk down a long hall to reach a back bedroom that is being used as an office, a lot of professionalism is lost.

I've seen, created, and remodeled a lot of home offices. I've seen enclosed porches turned into offices. Laundry rooms sometimes double as offices. My writing career started in a crowded laundry room. Bedrooms can be used as work spaces, and mega-buck businesses have been launched from dining room tables. Basements often are used as offices, and even the back wall of a living room might house business equipment and a makeshift office. There is no limit to the locations available for offices. But finding and creating an ideal work space is something that doesn't always come easily.

Garages

Garages perhaps are one of the most overlooked sites for ideal home offices. Due to their nature and location, garages can make great offices. Whether attached or detached, garages are far enough away from finished living space to provide a buffer of privacy, an important consideration for many home-based workers. Trying to work in a converted dining room while kids are running, jumping, and yelling in the adjoining living room can be all but impossible. When an office is in a garage, there is no need to deal with other family members while working. From my perspective, this is a key factor in choosing an office location.

Working from home is difficult for some people. For those without a lot of self-discipline, working from home can be anything but productive. This is never more true than when it's easy to walk into the kitchen, converse with a spouse from the work area, or interact with children in the next room. Entrepreneurs who can sequester themselves into a separate and private work space tend to get a lot more accomplished during their designated work hours. Garages provide the possibility for the seclusion needed to concentrate on work.

I've worked from home off and on for most of my career as a business owner. Even when I've maintained commercial space, I've preferred to work at home. Since my work frequently involves having deliveries made and customers dropping by, I feel most comfortable when my home office has a private entrance. I believe a separate entrance tells visitors that my office is in my home, but it is also a serious, professional work space. It also doesn't hurt to have a separate entrance when justifying home-office expenses to the tax collector.

Garages can be used on either ground level or, in some cases, in the attic area, to create a professional home office. I've done many types of garage conversions. On occasion, I've removed the existing roof of a garage and replaced it with a different style or pitch to make room for living space. Such conversions can be expensive, but there is a lot of square footage to be gained by converting a garage attic to living space.

Less expensive garage conversions that I've done for customers have been simple. I've gone inside the garage and created office space that is accessible from the side of the garage, through a standard entry door. This type of modification is completed quickly and is very cost-effective. The only downside is loss of parking space in the garage. Depending on the size of the garage, it can be possible to continue parking one or two cars in the space adjoining the new office walls.

I've known contractors who have converted their garages into combination office-warehouses. They build an office either in or over the garage and use the remaining garage space for storage and inventory. This type of setup is effective and not very expensive, compared to other options. Each business owner will have specific office needs. Once these are addressed, you can configure the most effective office plans. If a garage available, you should consider that option.

Building an office in a garage isn't difficult. You normally will be required to cut an entry door and some windows into the siding at some point. Beyond that, the rest of the job is typical. Framing is easy, and a ceiling structure is already in place. One of the room's walls will already be in place. The cost of this type of conversion is low, and the benefits are great. Aside from working with a concrete floor, the creation of a floor-level garage office is typical of any normal construction process.

Occasionally, the attic space of a garage is a good place for an office. Access to the upper level can be gained from inside or outside, with an interior access the more desirable choice. I remember a conversion on a garage attic where people coming to the office stepped through a standard entry door into a foyer. Steps went from the foyer to the space above. Since the stairs were attached to the interior of an outside wall, they robbed the garage's ground level of only about 3 ft in width. Roof design and pitch will, of course, have some bearing on how usable an existing garage attic is. In most cases, it will be necessary to increase the structural rating of what will become floor joists. Locating an office in the attic of a garage can be expensive, but it also can provide a very nice space. A real estate appraiser whom I deal with frequently moved out of her in-town office, which was in the same building where my in-town office had been, to new office space over her garage. Access to her office is gained by climbing outside steps, which I don't like. Once you enter the office, however, it is spacious, professional, and very comfortable. She and her husband both have desks in the space, not to mention tables, and a lot of computer gear, but there is still more than enough room to spare. Attic storage is available in the kneewalls of the office.

I have converted many garage attics into apartments for teenagers, in-laws, and paying tenants. With a little planning, it is possible to do a lot with the space over a garage. One carpenter I used to deal with had an office over his garage. The space was large enough to accommodate a full kitchen and bathroom, the office, and a private room for afternoon naps. He probably could have lived in the office had he wished. When your customers want a lot of square footage at a reasonable cost, a garage attic can be a perfect place for a conversion.

There are other advantages to using the attic space over a garage for an office. One of my favorites is the use of roof windows or skylights. My office is over a garage. The office was not an afterthought or add-on. When I built the house, I designed the plan myself. My office is accessed through a tiled foyer in the front of my home. When you enter the front door, you are in the foyer. Standing in the foyer, you can go through one of two metal entry doors. One leads into my living room, the other opens onto the stairs to the office. My office, which is large, is on the left at the top of the stairs.

In my case, I created more space that could have been fitted over the garage. I have a full bathroom, another office, and a third room that serves many purposes ranging from a photo darkroom, to storage, to a production area. This much space is excessive for most home offices. However, just the space over my garage could have been divided into two full-size offices and then some. My point is this: You can develop a lot of office space over a garage.

Basement offices

Basement offices are common in areas where homes typically have basements. Some of the office space is fairly primitive, but some is nice. During my career, I've run into all sorts of basement businesses. A general contractor for whom I once worked had his drafting table and office in his basement. The furnishings were fancy. My first book publisher ran a reasonably large publishing company out of his basement, and it was extremely professional in terms of staff, equipment, and appearance. Other types of offices that I've seen in basements have ranged from medical offices to drafting businesses. If a basement is handled properly, professional results can be obtained.

We've discussed the options for finishing basements, and which types of basements best lend themselves to living space. Those same factors apply for basement offices. A walk-out basement is best, and a daylight basement is second best. Buried basements are last on the list. All basements provide potential for home offices, but some more than others. I think we've discussed basements and access enough to skim over basement options at this point. We will, however, come back to basements when we get to the section on properly setting up a home office.

Attic offices

Attic offices probably are the least common of the types of home offices. Putting an office in an attic can be difficult from an access point of view. Not only that, attics are hot in the summer and cold in the winter, ruling out casual offices. To make an attic office comfortable, a person has to commit to a serious conversion project. There also are some design aspects of attic space that contractors should keep in mind when thinking of conversions.

Spiral stairs are sometimes used for space-saving access to converted attics. This is fine for some uses, but the limitations of the stairs make it difficult to move furniture up or down. Trying to get a desk up a spiral staircase is not an easy task. A standard set of stairs is needed to use attic space as a conventional office.

Many homes that have stairs leading into the attics are designed in a way that make them unsuitable for public offices. The attic stairs frequently originate in an upstairs bedroom. Not many business owners want clients and customers coming through a bedroom to get to an office. While this type of arrangement isn't suitable for a public office, it can be fine for other types of home offices. For example, my writing business doesn't require any face-to-face office meetings. I can work from anywhere in my house. Using bedroom stairs to access an office for this purpose would not create a problem. On the other hand, I wouldn't want my contracting customers to walk through my home to get to an office.

Many Cape Cod homes are built with the upstairs left unfinished. In this style of home, steps frequently lead to the upstairs from a hall. This type of arrangement can be used to create a public office. There is nothing wrong with putting office space in an attic, but the cost can be considerable. Roof pitch must be considered, and floor joists might have to be added to support the use of living space. If stairs are not in place, they must be installed. It can cost thousands upon thousands of dollars to get an attic ready for viable office use.

I can't remember many attic offices that were stand-alone rooms. I recall several homes in which attics were finished into living space which included an office, but I can't think of any houses in which the only finished attic space was an office. If you are working with an attic office, you have the luxury of using skylights and roof windows, which is, in my opinion, a plus.

Existing rooms

Conversion of existing rooms is probably the most popular method for creating in-home offices (Fig. 24-2). This is certainly the least expensive way to gain an office in a home. While it is cost-effective, converting existing rooms does not always work for public offices, but it can. Some house designs lend themselves well to the conversion of existing rooms for public use. If you are working with a house that has

24-2 *This conversion from a bedroom to an office was very successful, due to the cabinetry used.* _{Rutt, Lis King}

a room off a foyer or hall, that room can be considered a public office. By my definition, any room that can be reached without passing through other rooms qualifies for consideration as a public office.

It can be difficult to find rooms in a home that can be used as a public office. Unless a living room is converted, which is rarely practical, there might be no other room available that doesn't have to be accessed by passing through other rooms. Think about it. How many normal houses have multiple rooms that can be reached from a foyer or entry hall? Many houses have front doors that open right into living rooms or kitchens. The kitchen angle can be used to your customer's advantage if a public office is needed. Let me explain.

Assume that your customer wants a home office that is suitable for public contact. The house has a front door that opens directly into a living room. A side door opens into a breakfast area that adjoins an eat-in kitchen. These are the only exterior doors for the house, and none of the other rooms having outside walls are available for use. How can you get customers or clients into an office without parading them through other living space? Simple. Convert the dining area into an office. This is an easy conversion to make.

Install a partition wall with a door in it between the breakfast area and the kitchen. Once this wall is in place, the dining area becomes a separate room with independent access. Constructing one wall is all it takes to make a viable public office. Since the existing flooring is probably vinyl, you might want to cover it with carpet, but that's really the only conversion needed, and is not a necessity. As long as the house has a dining room or a large, eat-in kitchen, in addition to the breakfast area, your customer is in business in style.

Let's consider the same home, but assume that there is no other dining area available and that one bedroom is vacant. You could cut a new exterior door into the outside wall of the bedroom to create public access directly to the new bedroom-office. I've made this type of conversion on several occasions, and it's easy and inexpensive to do. You've just seen two examples for using existing finished space to make nice offices at nominal costs.

If public access isn't a concern, any interior room can be transformed into an office. It can be a bedroom, a laundry room, or a dining room. I even converted a walk-in closet into an office on one occasion. Surprisingly enough, the room turned out to be a nice, cozy office for a woman who did bookkeeping and payroll work out of her house. She met clients in her living room and did the real work in the old closet. The setup worked fine for her.

My first writing office was in a laundry room. As my success as a writer grew, I moved the office to a closed-in porch. From there, I took a large bedroom in the next house I lived in and made that my office. Now I have a substantial office suite in my home. During the time that I maintained an in-home office for writing, I had commercial office space for my other business interests. Now I run all of my businesses out of my home with a high degree of professionalism and a low overhead.

The requirements

Requirements for home offices vary from person to person and business to business. Some home workers don't need much more than a chair, a table, and a telephone. Computer stations are required in many home offices, and it's not unusual for more elaborate arrangements to be desirable. As a conversion and cosmetic contractor, you might be asked for advice in planning home offices. Do you know what to do if this happens? You have to talk with each customer and determine individual needs.

Based on historical data, I would say that most home offices have painted walls and either painted or textured ceilings. Some rooms are much fancier, but most aren't. Carpet tends to be the floor covering of choice. Good lighting is a must, and so is an abundant supply of electrical outlets. It simply doesn't do to have extension cords running from all corners of an office. Built-in bookcases are popular, as are built-in computer stations. In general, the cosmetic aspects of most home offices are simple in nature. You probably will run into some jobs in which flooring and wall coverings are upgraded, but this will be the exception, rather than the rule. Interior rooms can be converted easily to office space. If any remodeling work is needed, it usually is electrical work. Aside from providing additional outlets or separate circuits, the rest of the cosmetic work usually involves basic painting and flooring. The bulk of your responsibility with home offices will be to make sure that they are large enough and positioned properly to handle the needs of your customers.

25

Home exteriors

The exterior of a home, because it is the first thing people see, should make a good impression. Real estate brokers refer to the exterior's impact as *curb appeal*. A house that doesn't have good curb appeal can languish on the open market for a long time before being sold. Bad curb appeal can work against brokers and owners who want to sell houses. It also can be bad news for homeowners who have no intention of selling. Since home exteriors are seen by more people than are interiors, it stands to reason that property owners take an interest in the appearance of the outside of their homes. This is good for cosmetic contractors.

Homeowners who have ugly laundry rooms or basements might be willing to live with the deficiency for a long time. Those same homeowners might not be inclined to let the exteriors of their homes deteriorate. Many homeowners dress up the outside of their homes long before they pay any attention to interior problems. This means that cosmetic contractors can enjoy brisk business when they concentrate on exteriors.

When we talk about home exteriors, there are a lot of topics to consider (Fig. 25-1). Some of them, like decks, porches, and landscaping, will be covered in subsequent chapters. This leaves windows, doors, railings, roofing, siding, and walkways. Siding and roofing are the two most common cosmetic improvements pertaining to home exteriors. These two types of improvements are the ones that most contractors concentrate on, but that doesn't mean that they are the only ones to consider. In fact, some smaller aspects of exteriors can play large roles in the overall appearance of a home.

There are so many exterior improvements possible that it's difficult to decide where to start. House styles have something to do with what improvements should be considered. I went through a phase where it seemed like I was getting a run on colonial homes. The houses on which I worked all had concrete steps and stoops. None of the houses had roofs over the stoops, and this made the front elevation of the

Chapter Twenty-five

25-1 *A good example of what a home exterior should look like.* _{Vellux}

homes look a little awkward. All of the houses involved in my run were in the same subdivision, so I assume that after I did the first job and neighbors saw the improvement, they called me. I know this was the case with some of the jobs. Before we jump into specifics for all exteriors, let me share this little war story with you.

I did a cosmetic overhaul on a colonial home that included adding pillars and a covered porch over the front stoop. The addition of the A-roof over the porch made a huge difference in the appearance of the home. Even I was amazed at the improvement. Before the roof was added, the house looked too tall and stark. The new roof broke up the lines of the house and gave some much needed depth to the front of the home. After doing this job, my phone started ringing with new customers in the same subdivision. They all wanted roofs installed over their stoops.

If my memory serves me correctly, my crews altered four houses in nearly identical fashion. One or two of these little jobs led to much larger interior work. I guess I started a trend, or something. Getting into a subdivision on one job can lead to a lot more jobs, even if you don't aggressively solicit them. Neighbors talk, and job-site signs sell a lot of work. This is one big advantage to exterior improvements. Neighbors can see what you are doing and are likely to contract you for work on their homes.

Windows

I consider windows to be the eyes of a home. There is no doubt that windows have a lot to do with how a home looks. Replacing all of the windows in a home is an expensive proposition. This type of work is hard to justify, unless the existing windows are old and lack good insulating qualities. Many contractors don't consider doing partial window replacements and even fewer homeowners think about this tactic. I believe this is a mistake.

How many windows does the front of an average home contain? The number varies, of course. In some cases, the front windows can run into double digits, but a lot of houses don't have many windows. It might be possible to consider replacing all windows on the front of a house without your customers incurring unbearable costs.

I remember a ranch-style house on which I did exterior cosmetic work many years ago. The house was very plain, and didn't command attention from the street that ran in front of it. To change this, I replaced two living room windows with one large bow window. Small, double-hung windows dotted the two bedrooms that had outside walls on the front of the house. I replaced these common windows with larger windows that included half-moon transoms. A total of three new windows and a new front door were installed. After the work was completed, the house looked so different that people who had visited in the past had trouble locating it. They honestly didn't recognize it as the same house. Think about this. Three windows and a door change a house so much that guests who have visited before don't recognize it. If this isn't good advertising for affordable exterior cosmetic conversions, I don't know what is.

The size and shape of windows play an important role in the demeanor of a house. Tall, skinny windows look very different from short, wide windows. Something as simple as adding snap-in grids can make a noticeable difference in a window. Window replacement is not always easy, but it is rarely a major job. When you get your next call to change the personality of a house, consider doing something fresh with the windows.

Doors

Doors are a key element of a home exterior. I'm sure that if you think about the last few houses you went to, you will remember some of the doors. One of the worst doors I've ever replaced was peeling. I'm not talking about only the paint, which was nearly nonexistent. The door itself was coming apart and peeling. This was an old door that had three staggered pieces of glass in it. Gray, shingle-type siding surrounded the once-white door. Except for the door, the home's exterior wasn't in too bad shape. All of the windows needed reglazing and their trim needed to be painted, but the door was the real attention-getter. It looked terrible.

I had my crews fix the windows and replace the door with a six-panel, metal, insulated door. It was painted red, which might not have been my first choice, but the customer liked it. In the end, the red door looked good. Adding a splash of color in the middle of a gray mass made a favorable difference. On this job, the only major material investment was the door, and it cost less than $150, including hardware.

I have probably installed every type of exterior door made for modern construction. Steel insulated doors get the nod more often than not. Fiberglass doors have become popular because they can be stained like wood doors but don't swell in damp weather. Some customers swear by wood doors, and other people swear at them. In any event, wood remains a popular choice. One of the most popular doors I install has a metal, insulated lower half and a fancy glass panel in the upper half. There are many types of decorative designs available in these doors. I have this type of door on the front of my house.

An accessory to a front door is side lights. These, as you probably know, are narrow glass panels placed on each side of a door. Some side lights run from floor to ceiling and others are only half glass. If there is enough space around an existing door to install side lights, the effect can be dramatic. Don't expect this type of work to be cheap, but let your customers know it is an investment that will enhance their homes for years to come.

There is a multitude of entry doors that can be purchased for less than $300, and there is a good selection of doors available for half that cost. When you consider the low cost of doors, it should be obvious that doors are an excellent way to improve the exterior of a home with minimal cost.

Railings

Railings for steps, stoops, and decks are so common that they tend to be taken for granted. It's not unusual for builders to install low-cost railings on expensive homes. I feel that's a mistake. I'm not suggesting that you install gold-plated railings, but there is no justification in cheapening the appearance of a home to save $100 on a railing.

When you evaluate the exterior of a home for a cosmetic conversion, take a close look at the railings. Some types of railings scream out old age. Many railings become weak and wobbly, and should be replaced for safety reasons, as well as for appearance. When you look at railings, pay attention to them. Are you dealing with an attractive home that is fitted with rough-looking railings? Would the house benefit from a more polished style of railing? The chances are good that it would.

Many of the houses I see have standard, simple, pressure-treated handrails and square pickets. This arrangement is okay on some houses, but it looks out of place on many house styles. Pressure-treated wood doesn't take paint well. Having greenish bare wood on the front of a fancy house can be a detriment to good curb appeal. Some type of carved handrail and turned balusters that are painted offer a more formal look. This type of wood costs considerably more, but not a lot of it is needed.

Roofing

Very few homeowners think of roofing when it comes to exterior home improvements. About the only time the roof comes up is when it's leaking. Yet changing the roofing on a house can make a big difference in the home's appearance. Something as simple as going from a black roof to a white roof can create a very noticeable difference. Color is not the only consideration. Changing the shingle style can make even more of a difference.

Regular asphalt shingles account for most of the roofing on modern homes. Slate roofing can still be found on older homes and very expensive properties, but it is becoming something of a rarity. Tile roofs show up on some house styles, but are not found in most neighborhoods. Dimensional shingles add depth to a roof and set it apart from others. Cedar shakes often look good when they are installed and look bad after several years of use. Let me give you an example of how cedar shingles, which cost more to install, can become a big problem for homeowners down the road. My parents live in an upscale neighborhood. Their house has asphalt shingles that are showing their age and should be replaced soon. Several of their neighbors have cedar shake roofs. While it's obvious to a trained eye that the shingles on the roof of my parents' house are getting close to replacement, the cedar roofs on adjacent houses have turned into eyesores that anyone could tell should be replaced. We're not talking about cheap houses here. The cedar roofs cost extra when they were installed, and now they are falling apart, turning green, and looking old. I don't know if they are leaking, but they rob the homes they cover of good curb appeal.

I don't know about you, but I'd be one upset homeowner if a roof for which I paid extra looked worse than the less expensive roofs on the homes of my neighbors. If you consider your business to be a long-term investment, you should give some thought to this type of situation. People for whom you install cedar roofs might be very happy, but how will they feel about you and your company 10 years from now? I'm not trying to bad-mouth cedar roofs. I think they look good, until they don't.

I've repaired slate roofs, tile roofs, and cedar roofs, but I've never acted as a contractor to install any of these roof types. My work has been limited to asphalt and fiberglass shingles. These roofing materials are so good that I have not needed to work with other roofing materials. Fiberglass shingles have taken over my local market in recent years, but I still prefer the time-tested asphalt shingles. You and your customers might feel differently.

Regardless of what type of roofing you are replacing or installing, you should consider roofing to be an element of cosmetic contracting. Some houses simply have ugly roofs. I mean, a green and black roof, where does that come from? But, I've seen them. Take a look at the roofing on houses for which exterior improvements are planned. Even if the roofing isn't in bad shape, your customers might be willing to replace it to gain a better appearance.

Siding

Siding captures most of the attention of people assessing the exterior of the house. This is logical, since siding accounts for most of a home's exterior. There are a number of types of siding in use today. I find the most frequently used materials are hardboard, pine, cedar, and vinyl. These selections are based on new construction. When you look at older homes, you find other types of siding, such as aluminum, and shingles, which often contain asbestos.

Most contractors have personal preferences when it comes to siding. My universal favorite is pine siding that is stained, although I've had subcontractors install hundreds of squares of hardboard siding, and plenty of vinyl siding. Every home I've built for myself has had pine siding. Why? Because it's relatively inexpensive, it looks good, and it wears well. And I like the look.

Contractors have to be careful not to let their personal preferences get in the way when advising customers on what materials to select. If you think vinyl siding is wonderful, that's fine. But, you shouldn't shove vinyl siding at your customers like it's the only material available. To be fair to customers, you must present them with options.

Vinyl siding Vinyl siding is popular with some homeowners. The fact that the siding never has to be painted is a big selling point, and this is a good feature. When contractors pitch the siding as being maintenance-free, however, I have a problem. Vinyl siding doesn't require painting, but it is not unusual for the siding to require power washing to remove dirt and mildew. This is not what I would call maintenance-free.

Plenty of homeowners are willing to have their homes clad in vinyl siding to get a new look without the ongoing duty of painting. There's nothing wrong with this, but they should be advised that the siding might require periodic cleanings. I feel that contractors who don't offer this information are taking advantage of uninformed homeowners, which is something that I detest and that taints the general contracting business.

Vinyl siding does offer a variety of benefits. It's durable, doesn't fade appreciably, and never needs painting. These are all strong selling points. But vinyl siding is considered by some to be less valuable than wood siding, and therefore can reduce the appraised value of a home. Can you imagine paying a contractor to improve your home, only to find out that the value of your house decreased even with the expense of new siding? This would put me into some sort of rage. I might add that this is not just my opinion. I interviewed real estate professionals, such as appraisers, and found this scenario to be true.

Aluminum siding Aluminum siding is all but a thing of the past in the areas where I work. Another author told me recently that aluminum siding is still popular in some areas. The place he mentioned by name was Illinois. Personally, I can see no reason to use aluminum siding when vinyl siding is available. Homeowners sometimes have aluminum siding replaced because of dents, fading, and scratches. I don't doubt that you will replace aluminum siding during your cosmetic construction, but I don't believe you will install much of it as a siding replacement.

Shingle siding Shingle siding clearly dates a home. When you see a house with this type of siding, you can count on it being old. By shingle siding, I'm talking about the large squares used as shingles for siding, not cedar shakes. Some new houses are still being built with cedar-shake siding.

If you are hired to replace shingle siding, be careful. Much of this type of material is known to contain asbestos. You probably will be better off installing new siding over the old instead of removing the shingle siding. Check with authorities in your local jurisdiction to see what requirements pertain to this type of siding.

Cedar siding Cedar siding, in clapboard form, is a popular siding material. It's expensive, but builders use a lot of it. Cedar is known for its durability. It is durable, but if you don't stain, seal, or paint cedar siding in short order, it will begin to weather. This usually means that it will turn gray. Some homeowners like this. As a cosmetic contractor, you probably won't remove much cedar siding for replacement. Instead, you probably will simply paint or stain it.

Pine siding Pine siding is what I use on many of the new homes I build. Once the siding is installed and painted, it's difficult to tell from cedar. Staining pine siding brings out the wood's grain and knots, and this can be appealing. If pine siding is not sealed, painted, or stained promptly, it will begin to turn black. As with cedar siding, I doubt if you will do many removals of pine siding for some type of replacement. The last time I checked, pine siding cost much less than cedar and about the same as high-quality vinyl.

Hardboard siding Hardboard siding is affordable and looks fancy once it's painted, if it is installed correctly. This type of siding is flexible, and can warp into uneven shapes. This is particularly true if the siding is hauled around on the rack of a truck, or is not fully supported when stored. Hardboard siding requires effort on the part of an installer to maintain straight lines. If this material is handled, stored, and installed properly, it looks great. The only downside to hardboard siding is that it will not take stain; it must be painted.

Other types of siding There are other types of siding. For instance, there is a type of siding that comes in 4×8 sheets that is manufactured to look something like shiplap siding. Many homeowners use this type of siding on storage buildings and add-on garages. I used it on my first house, but wasn't all that pleased with it. It's easy to install and fairly inexpensive, but it doesn't give a house a good exterior finish. This may be the type of siding that you will be asked to replace or cover up.

In addition to the 4×8 siding panels, you can find houses with other types of siding. Boards might be used to create a herringbone or shiplap pattern. Stone might be in place as an existing siding. Brick is still used for the exterior walls of some homes, and there are other options. Most siding work involves painting and staining, with some occasional replacements.

Gutters

Gutters are not installed on all houses. Many homeowners feel that gutters detract from the appearance of a home. While gutters are not beautiful, they are very practical. Old gutters that are sagging and falling off their mountings are particularly ugly (Fig. 25-2). It is this type of gutter that might make your phone ring. I don't think many homeowners call contractors to have gutters installed for cosmetic reasons, but they do call to have old gutters removed or replaced. This is simple work that can be done quickly. You won't get rich off gutter jobs, but they can lead to bigger and better work (Fig. 25-3).

Material	Expected Life Span
Aluminum	15 to 20 years
Vinyl	Indefinite
Steel	Less than 10 years
Copper	50 years
Wood	10 to 15 years

All estimated life spans depend on installation procedure, maintenance, and climatic conditions.

25-2 *Potential life spans of gutters.*

Material	Price Range
Aluminum	Moderate
Vinyl	Expensive
Steel	Inexpensive
Copper	Very expensive
Wood	Moderate to expensive

All estimated life spans depend on installation procedure, maintenance, and climatic conditions.

25-3 *Price ranges of gutters.*

Shutters

Shutters are an external improvement that I never used to think much about. They aren't functional, so why install them? Looks, that's why. Adding shutters can make a big difference in the looks of a house. It took me a long time to realize just how big a difference shutters can make.

I built a house on speculation that didn't have shutters. A young couple saw the house after it had been sold and wanted me to build them one just like it, except they wanted shutters on their home. I didn't give much thought to the shutters. The house was built and the shutters were installed. I took a picture of the house to put in my photo album. When I inserted the picture in the album, it happened to be opposite a picture I had taken of the spec house that didn't have shutters. Seeing the two houses side by side made it clear that shutters do make a difference. Since that revelation, I've installed a lot of shutters on homes.

When you consider shutters are available that never have to be painted and are pretty cheap, it makes sense to use them. Not all houses look better with shutters, but many do. Adding shutters to an existing home doesn't take long or cost much, but it sure can change the look of a house.

Lighting

Lighting can have a lot to do with how nice the exterior of a home appears. A lack of lighting does nothing to improve the looks of a house, and old lights can detract from appearance. New lights that showcase a home can produce tremendous effects. Something as simple as replacing a black, plastic wall lamp near a front door with an antique brass lamp can be all it takes to make a difference. The fixture is affordable, and the labor required for this type of replacement is minimal.

Light fixtures around entry doors are the primary electrical considerations involved in most exterior electrical work. Installation of low-level floodlighting that shines on a house, or walkway lights that produce a soft glow, can work wonders. Light fixtures and lighting might not come quickly to mind when you think of exterior improvements, but electrical work should be given full consideration.

Porches and decks could be considered part of a home exterior. But, since these two topics attract a lot of attention on their own, I think they deserve their own chapter. In the next chapter, I will begin to cover the cosmetic opportunities for decks and porches.

26
Decks and porches

Decks and porches offer cosmetic contractors two options: building new decks and porches, or sprucing up existing ones. The addition of outside living area not only makes a home more enjoyable, it is a way to change the looks of a house. Taking a rundown porch and putting it into tip-top shape can improve the curb appeal of a house. Making an old deck look flashy again also can work wonders for the outside appearance of a home.

I've done a lot of work with both porches and decks. Of the two, decks have been the more common work in my area. There are, however, many types of porches that need repairs and refinishing. As a cosmetic contractor, you can benefit from all aspects of porches and decks. Let's start our discussion with existing porches.

Existing porches

Existing porches take many forms, ranging from screened porches to those that are little more than covered slabs of concrete where people gather during cookouts. Then there is the lovable, old-fashioned front porch. You know the type I'm talking about. It runs the full width of a house and might even wrap around one or both sides of the home. Large columns support the roof structure that protects the concrete floor. This type of porch is popular in the South, where I grew up.

My grandparents had a large front porch, and some of my fondest memories were formed in the shade of that porch. I shelled beans and peas while rocking in the glider on that porch. My grandfather and I played countless games of checkers as we watched summer rains splashing in the puddles beyond the security of the dry porch. All sorts of other games were played under the shelter of that big porch, and the highlight of summer days was swaying in the porch swing, eating blackberries, while my grandfather and I eagerly awaited the mail delivery. There is a lot to be said for the benefits of a covered porch.

Screened porches are popular in the South. I've made thousands upon thousands of dollars building new screened porches, and I've done pretty well repairing older ones. Screened porches generally are built with less than the best of materials, and this leads to a lot of opportunity for cosmetic contractors.

Front porches on older homes typically have concrete floors, wood ceilings, and large columns that support the roof structure. This type of porch seldom requires more than an occasional coat of paint to keep it looking good. Sometimes, however, the columns begin to rot away and must be replaced. This is an opening for cosmetic contractors. If the ceiling begins to go bad, it can be replaced with new wood or moisture-resistant drywall. This is yet another opportunity for cosmetic contractors. Some homeowners tire of looking at a gray concrete floor and opt to have indoor-outdoor carpeting installed over the concrete. This also is a possibility for contractors to pick up on.

Newer porches often are made with wood floors instead of concrete. When wood is used, it might need painting every few years. Railings can be repainted or replaced with more decorative designs. Sometimes floorboards swell, settle, or rot. Replacing these wood members is not a lot of work, but it opens the door for other jobs, and it keeps a porch safe and looking good.

In addition to routine maintenance, some front porches are large enough to benefit from add-ons, such as additional lighting or ceiling fans. It could even make sense, in some cases, to consider screening in a porch. Little touches, like a ceiling fan, can add considerably to the enjoyment and appearance of a porch.

Adding a porch

Adding a porch to a house can make a huge difference in the home's appearance. The entire look of a house is changed when a front porch is built. I've looked at thousands of house plans during my career. In doing so, I've gotten a good feel for how a house can look very different depending upon it's roof design and slope, and whether or not a porch is installed. Let me give you a quick example of what I'm talking about.

I remember clearly a house that one of my competitors used to build in several subdivisions. The house was basically a ranch-style home that had a tall roof, making it look more like a Cape Cod. Inside the home, a loft took advantage of the steep roof pitch. From the road, this house design looked awkward. The roof pitch was out of proportion to the house, or something. It's hard to explain what it was about the house that didn't look right, but it just didn't. Apparently, my feelings were shared by the general public. The first home of this style that was built stayed on the market much longer than other homes built on adjoining lots. Inside, the house was nice, but it lacked curb appeal.

After the house had been on the market for a considerable length of time, the builders decided to modify it. They did this by building a front porch on the house. The porch roof tied into the main roof, breaking up the long, stark appearance. Since the house was rustic in nature, the front porch was built with wood posts and floorboards. The house sold in less than 2 weeks after the porch was added. From that day on, I never saw those builders put up that house without building the porch. I can't describe how much difference the porch made. With a porch, the house design reminded me of an old-fashioned mountain cabin.

Spec builders, as you probably know, have to be careful about what they build and where they invest their building money. A porch is not always considered a good option in terms of spec building. I broke this rule once, and was I ever glad I did. The spec house I was building was a large farmhouse design. The blueprints showed a wraparound front porch, but I almost deleted it. At the last minute, I decided to go ahead with the porch. After completing the house, I was glad that I had not eliminated the porch. This house was used as a model home, and I got numerous compliments and comments on what a wonderful design I had built. The porch made the difference.

The two stories you have just read pertain to new houses, but the information also applies to existing homes. The addition of a porch to a bland house can put a lot of zip in the curb appeal. Even if the porch is no more than a stoop cover, it can improve the exterior of a home tremendously. I've seen many colonial homes with and without stoop covers, and I can tell you that the little front porch makes all the difference in the world.

Front porches are expensive to build, but they are sometimes worth it. I stress the word "sometimes." Many appraisers don't feel that porches are worth what they cost in terms of resale value. This may or may not be true. It depends on comparable sales in the area. What cannot be denied is that front porches have the power to make a house look different, in a good way.

If you propose a porch to a customer, someone must decide on the nature of the porch. Will it be formal or rustic? Are you going to use pressure-treated lumber or wood that will take paint sooner and better? These are just some of the considerations. Front porches can get quite formal, and this is appropriate for some styles of homes. Many houses, however, do just as well, probably better, with a more casual style of porch.

You don't have to go to the expense of a poured foundation and a concrete floor. Pressure-treated lumber on a pier foundation that is enclosed with lattice works with most homes. If customers get carried away, they can spend as much for railing materials as some people would spend for a complete framing package. Porches offer homeowners and contractors a variety of options, so give this type of work some thought the next time you are stumped for a good idea that will change the front elevation of a house.

Existing decks

Existing decks can get in bad shape fast. This is especially true of decks that were not built with pressure-treated material. The damage can be structural or cosmetic. In either case, there is work for a contractor. With decks being so abundant, they offer plenty of work for enterprising contractors.

Cosmetically, existing decks are difficult to deal with. Replacement boards for pressure-treated decks never match in color with the old boards. Painted decks can be scraped down and repainted, and this is a good chance for improvement. You could remove existing railing and replace it with a more attractive style. Changing the steps on a deck might improve the physical appearance. On the whole, though, it can be tough to make major improvements on existing decks. There is an exception to this statement.

One way to make the most of an existing deck that is in bad shape is to strip the deck down to the floor joists. You can salvage the foundation and floor framing. You then can build a new deck on the old base, which makes economic sense. Anything short of this is probably going to look like a patch job in cases of pressure-treated decks. Painted decks are easier to improve, but I haven't seen a lot of painted decks.

The railing used on a deck affects the overall appearance of the outside living area. Many carpenters use simple pickets in a vertical format to meet code requirements. I've done this with a lot of decks. Vertical pickets aren't noteworthy. They don't capture attention. Getting a little creative with patterns in pickets can overcome this.

There is a deck that I drive by from time to time that really catches my eye. The railing is what does it. Someone took the time to build the railing with diamond shapes in it. There are vertical pickets in the railing, but they serve as anchor points for the diamond framing. The railing is unlike any other I've seen, and I can't help but look at it whenever I'm passing by. This is a prime example of how a little ingenuity and some extra time can take an ordinary deck and make it extraordinary.

New decks

New decks give contractors a chance to really shine in what they do (Fig. 26-1). You can build one-level decks, two-level decks, three-level decks, decks with built-in spas, decks that terrace out over an entire lawn area, and a lot of other types of decks. Given the flexibility in building decks, a contractor should never be at a loss for solutions in this type of improvement. Let me tell you about a theme deck I built for a customer.

A retired naval officer had a rather unusual house that was in the shape of a U. This man liked to entertain, and he was fond of his memories at sea. When he asked me for advice on a deck design, a particular design popped into my head almost immediately. It was perfect for this homeowner.

The deck I proposed covered all of the ground within the U-shape of the house. Where the deck met the end of the house, I suggested it be cut on a 45-degree angle at each end, creating a shape similar to a ship. Lattice was used to enclose the only open end of the pier foundation. A large drink rail was installed on three sides of the deck. A captain's wheel was created and mounted at the far end of the entertainment platform. Colored spot and flood lights were installed in the soffits of the house to illuminate the deck. My carpenters installed built-in benches for seating along two sides of the deck. When we were done, the deck had a seagoing charm that pleased the customer to no end. This exterior improvement couldn't be seen from the road, but it certainly added a lot of spice to the house.

A woman who handled my advertising account called me for a price on a small deck. I went to her house and looked over the situation. The building lot sloped considerably, making use of the back lawn difficult, at best. The advertising agent explained that she and her husband enjoyed having guests over, and that she often held semi-business gatherings at her home. The house was nice, but this customer wanted an outdoor area in which people could congregate. Due to the slope of the lot, a backyard cookout was pretty much out of the question. A grill on wheels would roll away, and visitors would need to have one leg longer than the other just to stand on the slope.

26-1 *When you are given the opportunity to build a new deck, your options are unlimited.* Velux

Decks and porches

The original request on this job was a simple deck that would come off the living room and run along the back wall of the house. Nothing fancy, just an elevated platform that would provide level footing. Always thinking as a salesman, I began to pitch the idea of a multilevel, terraced deck that would make use of most of the rear lawn. I suggested built-in seating, flower boxes, a drink rail, and other custom features. My idea was going to cost a lot more than a simple deck. This could have been a valid objection, but I pointed out how useless the existing lawn was and how beneficial the new deck would be for family members, and how the deck could serve as a sales arena for the advertising executive. Once I mentioned the business angle of the deck, I had a commitment from the property owner. The deck turned out great.

You could spend days reading about decks. Entire books can be devoted to decks alone. And decks traditionally are a good investment in terms of resale value. Some decks are visible from a public view, but many are secluded. The fact that an exterior improvement can't be seen by passing motorists doesn't mean that the improvement is not worthwhile. Even if the homeowner is the only person to see or use the deck, it can still be a valuable addition to a home.

One big question pertaining to decks relates to the best material to use when building them. I believe that more decks are built with pressure-treated lumber than any other material. I know this is the case in the areas where I have lived and worked. Redwood is a popular material for decks at exclusive homes. Regular framing lumber also can be used. The disadvantage to building a deck out of pine or fir is that is might not last as long as a deck constructed out of pressure-treated wood or redwood. But pine and fir can be painted nicely, and pressure-treated lumber can't be painted very well for a couple of years. This fact alone can be a deciding factor for some homeowners.

When I build a deck, I use nosed decking for my floors. Not all builders do this. Many decks are built with 2×4 material used as a flooring. In my opinion, this type of flooring looks amateurish and cheap. Railings are another area where some builders cut corners. I use 2×6 stock for top rails, but some builders use 2×4 material. It's a matter of preference and cost, but I think the wider rail looks better and serves nicely as a drink rail.

The components of decks can be kept very simple, or they can be pushed to great lengths. Railings are a good example of this. A cheap railing can be made with 2×4 rails at the top and bottom, with 2×2 pickets nailed between them. Expensive railings can be given considerable definition by using stylish wood in all areas. Depending upon the amount of detail carved into the wood, and the type of wood used, an outrageous amount of money can be spent just on the railing materials.

Some contractors install only one set of steps for a deck. When conditions allow for it, I install two sets, one at each end. While penny-pinchers build narrow steps that only meet minimum code requirements, I typically build wide steps with nice handrails on both sides. Every builder has a particular style, and you will have to find your own.

Foundations for decks are almost always piers. The area under a deck on a pier foundation can be left open. Some people store items like lawnmowers and grills under tall decks. I normally suggest enclosing the foundation of a deck with pressure-treated lattice. I think the job looks better when this is done. In answer to storage objections, I have built hinged lattice doors to provide for storage under decks with enclosed foundations.

I often build in little extras with my decks. These might include benches, flower boxes, lattice for plants to climb, bird feeders, or tables to accommodate the needs of an outdoor chef during a cookout. You are limited only by your imagination when building decks. There is good money to be made in building decks, and many homeowners are pleased with the idea of gaining some outside recreational space. This is an excellent way for you to pick up some extra work while improving the exterior of homes and making homeowners happy.

The final chapter deals with landscaping. Of all the exterior improvements you can offer a homeowner, landscaping might be the most important. Roofing and siding are critical elements of a home's appearance, but never underestimate the power of landscaping. We will discuss the use of walkways, shrubbery, fish ponds, and other landscaping tools to create fantastic home exteriors.

27

Landscaping

Landscaping is an aspect of cosmetic improvements that far too many contractors underestimate. Studies and statistics prove that landscaping, in moderation, is a good investment. There are practical reasons for planting trees and shrubs, as well as the obvious aesthetic reasons. Landscaping isn't limited to living plants. Walkways, fountains, benches, garden pools, and other types of improvements can be considered landscaping.

When I first started building spec houses, I skimped on landscaping. It took me a few years to realize that by saving a few hundred dollars in production cost, I was hurting my chances for a quick sale. As I became a more knowledgeable builder, I started investing more money in landscaping. Can I tell you that this investment sold my houses faster? I can't prove it, but I believe it did. People inspecting the homes often commented on the beautiful flower beds, the little walkways with park benches, and other amenities. It's logical to assume that these items helped facilitate faster sales.

If you're a pure cosmetic contractor or remodeler, you won't be concerned about landscaping as an aid to selling houses that you build. But you can use landscaping to sell more jobs. If you make potential customers aware of their landscaping options, and give them an idea of what to expect from a finished job, you should be able to create more business for yourself.

Contractors might wonder why a homeowner would call a contractor instead of a landscaping company for landscaping improvements. People who already have their minds made up to do landscaping might very well call a landscaping company first. However, a lot of people don't give serious thought to landscaping improvements until they are pitched on the idea. And don't forget, some types of landscaping work are usually associated with contractors. Walkways are an example of this type of work.

Let's assume that you have been called to a house to estimate the cost of a new paint job. You can give a painting estimate and be done with it all, or you can pitch some landscaping ideas while you're at the home. Contractors who grasp opportunities like this are the ones who get more work. You could more than double the profit you would make painting by selling the customer on landscaping to go with the new paint job.

Landscaping covers a lot of ground (no pun intended). You could plant trees, shrubs, or flowers. The job might require the removal of plants that have grown too large. When foundation shrubs are planted too close to a home, they can cause moisture problems for the house. Large, dense foliage can pose a security problem by giving burglars a place to hide. Most landscaping is done to improve the public view of a property. Many homeowners, however, invest thousands of dollars to create private retreats with their landscaping. Entire books are devoted to the subject of landscaping. With so many possibilities, we can hit only the high spots. I suggest that you give some serious thought to what you are about to read and consider taking on more landscaping work, even if you subcontract the work out to professional landscaping companies.

Landscaping plans

Landscaping plans can be difficult for people to envision when standing in a yard and pointing to locations. If you tell customers that their new roof will be a rich brown color, instead of the tattered, old, black roof that is on their home, they can get a sense of what the roof will look like. The only image they must conjure is the new color. This type of logic applies to painting siding and other simple improvements. There is a difference in visualizing a color modification and in seeing an entirely new layout.

Many contractors are taking advantage of technology to make sales. You can take a notebook computer to a customer's house, hook it up to the homeowner's television, and give an impressive presentation with vivid colors and graphics. If you know how to use computer-aided drawing (CAD) programs, you can design anything from a new kitchen layout to a complete landscaping plan. Ready-made clip art makes the insertion of various types of trees and shrubs easy. You can even use 3D programs to give customers a better idea of what their finished job will look like. If you are willing to go to this type of effort, you can count on some additional sales. Just the fact that you are so professional in your presentation will sell some jobs that might have gone to other contractors.

Not all contractors like using computers. With the direction technology is taking businesses, becoming computer literate is almost a necessity. If you refuse to go online, at least show your customers some type of rendering for what you have in mind. A few simple drawing tools and some graph paper will enable you to do this. One way or the other, be prepared to show customers what they will get when you are selling landscaping.

Trees

Trees are often planted during new construction. They are also used in cosmetic landscaping for older homes. A row of Eastern white pines, which grow quickly in most soils, can be planted to create a privacy screen. Paper birch trees are popular for landscaping because of their white, paper-like bark and their hardiness. Norway maples are used in cities and they grow fast. Dwarf trees are common in decorative landscaping. Before you start advising customers on what types of trees to plant, get to know your product line. Many trees get too large to plant in close proximity to a house.

Foundation shrubs

Foundation shrubs are an excellent landscaping tool. These low-growing, bushy plants can quickly mature to the point that they conceal foundations. If you don't think landscaping is important, look at a house that has no foundation shrubbery and compare the appearance to that of a house that has a good screen of shrubbery. There's no comparison.

As with trees, dozens of types of foundation shrubs are available. Each has its advantages and disadvantages. Some grow tall, and others stay close to the ground. Many retain their green color throughout the year, while others lose their color or their leaves. A quick trip to your local library or nursery will get you up to speed on the various types of foundation shrubs.

Flowers and groundcovers

Flowers and groundcovers are appreciated most for their appearance. Planting groundcover, however, is an excellent way to control erosion around homes that are built on sloping lots. Landscaping can be used as both a practical and pretty improvement for a home. For example, climbing cover, such as ivy, can be trained to grow up latticework, closing in the underside of a deck that is skirted with pressure-treated lattice. Explaining little facts like this to your customers will give you an edge in competitive bid situations and should result in more sales.

Timbers and mulch

Landscaping timbers and mulch can improve the appearance of a home. Timbers can be used to surround trees, making grass cutting and trimming easier. They also can be used to create terraces and other arrangements, such as planting beds. Let me tell you a quick story about how timbers, mulch, flowers, and groundcover were used to solve a serious problem for one homeowner.

The house in question had a small back yard, which was surrounded on three sides by a large hill. When I first saw the hill, it had scraggly brush near the top and the face had eroded. The homeowner complained about having miniature mudslides that filled in the small, but nice, lawn. I wasn't at the house as a contractor, but I made some suggestions on how the problem could be solved. The homeowner listened intently and thanked me.

When I returned to the house with the hill on another occasion, I was surprised to see that my advice had been taken. The hill had been terraced with timbers. Flowers, shrubs, and groundcover had been planted in the terraces. What had once been an eyesore was now a beautiful asset to the home. Mud no longer slid onto the lawn. Terracing the hill and planting the soil halted erosion. This was an expensive fix, but it worked very well. I wish I could take credit for the good work, but the idea is all I can claim ownership of.

Walkways

It's surprising how many homes don't have walkways. A walkway makes a home more inviting, and it helps to keep visitors' feet dry and clean. Many materials can be used to build walkways. One of the simplest is walkway stones. These are easy to install and inexpensive, but they often sink and sag. Concrete walks have been used for years, and they are a dependable, although expensive, option.

Bricks set in sand can create a nice walkway. Colored pebbles make interesting paths, and pressure-treated lumber can be used to make a walkway. Slate and flagstone also can be used. There is no shortage of walkway materials. Adding a walkway to a home improves appearance and access.

Garden settings

Garden settings can be relaxing. Contractors who design and build residential gardens often are in great demand. It's not uncommon to do a job for a customer and have neighbors calling for quotes to install gardens at their homes. This has proved to be the case with my company on many occasions. When you get involved in building backyard gardens, you can get into some extensive work.

Home gardens vary in size and degree of complexity. Some are little more than a few flower beds, a bird feeder, and a park bench. Other are expansive and expensive. I'm familiar with a garden that covers about half an acre of ground owned by a busy executive. The owner goes into the garden and shapes the trees and plants as a form of stress reduction. The garden is breathtaking. Many of the plants are common varieties, but they have been trimmed and shaped into unique pieces of art.

I've had success with creating water gardens for homeowners. Installing an in-ground pool is easy with the modern membrane materials available for use as liners. Putting a fish pool in a garden is a tremendous addition to the space. Couple this with some aquatic plants, proper lighting, and surrounding plants, and you have a retreat that anyone could appreciate. Creative cosmetic contractors can work wonders with the grounds of a home.

Sprinkler systems

Underground sprinkler systems have become popular with residential customers in many parts of the country. Lawns that burn and turn brown in harsh summer heat are not attractive. Many property owners working long hours at their jobs simply don't have the time to water their lawns. What was once a relaxing chore has become a task that there is no time to fulfill. This has served as a catalyst for the installation of automated sprinkler systems.

There are companies that do nothing but lawn irrigation work. Many plumbers are adding lawn sprinklers to their lists of services. This doesn't mean that you can't grab a piece of the action. If you are in front of homeowners selling them your services as a contractor, it's fairly easy to add to the list of services you provide, such as the installation of underground sprinkler systems. People generally like to deal with one contractor for all their needs. If you have a stable of subcontractors standing by, you can give customers what they need with one-stop shopping.

Many types of sprinkler systems are available. There are sprinklers for shrubs, sprinklers for trees, sprinklers for flower beds, and sprinklers for lawns. Different sprinkler heads can be connected to a common sprinkler system and controlled either manually or automatically. The price for installing a sprinkler system can be steep, but many homeowners are willing to pay it. Don't overlook this add-on sales possibility when discussing landscaping with your customers.

You probably are not a landscaping contractor and you might have no desire to specialize in landscaping. In that case, you're like most general contractors. But this doesn't mean you can't capture more of the home-improvement market by expanding your services. Remember, once you are in a customer's home, you are the salesperson of the moment. If you do your sales job right, you will get the work. The more you can offer customers, the more money you stand to make. Venturing into landscaping might seem like a long stretch, but it's not as difficult as you might believe. At the very least, consider offering your customers landscaping options when the next appropriate job comes along.

Index

A

Acrylic paint, 16, 47
Air conditioning (*see* HVAC systems)
Alkyd paint, 16, 47, 48
Aluminum siding, 11–12, 299
Appliances
   dishwashers, 81
   garbage disposals, 80–81
   kitchen, 183–184
Asphalt shingles, 15
Attics, home offices in, 288–289
Awning windows, 131

B

Barn board, weathered, 27–28, 223
Baseboard units, 111
Basements, 245–252
   bedrooms in, 246
   ceilings, 220, 222, 249
   doors, 251–252
   family rooms in, 219–222, 247
   flooring, 248
   home offices in, 288
   laundry rooms in, 269, 271
   lighting, 222
   upgrading, 245–247
   walls, 221–222, 249–250
   windows, 250–251
Bathrooms, 187–204
   accessories, 164–166, 200–204
      medicine cabinets, 165, 202–204
      mirrors, 201
      towel holder, 202
   bathing units, 92–94, 196
   bathtubs, 82–83
      clawfoot, 82
      whirlpool, 83
   cabinets, 150
      medicine, 165, 202–204

Bathrooms (*cont.*)
   ceilings, 197
   doors, 198–200
   fixtures, 197–198
      spacing requirements, 86
   flooring, 187–190
   lavatories, 85–89
      drop-in, 86
      faucets for, 90–92
      integral tops, 87–89
      pedestal, 87
      rim-type, 86
      wall-hung, 86
   lighting, 98, 198
   over-toilet organizers, 166
   plumbing, 75, 81–94
   tile flooring, 68
   toilets, 84–85
      improper alignment, 88
      width requirements, 87
      proper alignment, 89
   vanities, 151
   vertical storage units, 165
   wall treatments, 191–197
   walls in, 40–41
   windows, 198–200
Bathtubs, 82–83
   bathing units, 92–94, 196
   clawfoot, 82
   whirlpool, 83
Bay windows, 131
   walk-out, 264–265
Bedrooms
   accessories, 167–170
   adult, 235–243
      built-in cabinets, 242
      built-in fish tanks, 237–238
      ceiling fans, 236
      ceilings, 238–239
      doors, 242

Bedrooms, adult (*cont.*)
    fireplaces, 235–236
    flooring, 239–240
    lighting, 243
    skylights, 236
    terrace doors, 236
    wall treatments, 240–242
    windows, 238
  basement, 246
  child, 227–234
    built-in units, 232–234
    ceilings, 230–231
    flooring, 230
    wall treatments, 231
Block (*see* Cinder block)
Bow windows, 131
Brass kickplates, 161–162
Brick flooring, 55, 68–69
Brick foundations, 6–8
  bad bricks, 8
  bricks pointing up, 7–8
  cleaning, 8
  cracks in, 6–7
  soil loads, 7
Bricks
  bad, 8
  cleaning, 8
  on pier foundations, 4
  using for interior walls, 26–27

C

Cabinets
  bathroom, 150
  built-in systems for
    child's bedroom, 232–234
    adult's bedroom, 242
  choosing, 154–157
  construction features, 151–154
  costs, 146, 154–157
  features, 145
  installing, 157–158
    damage resulting from, 158–159
  kitchen, 143–146, 180–181
  lighting under, 163
  medicine, 165, 202–204
  refacing, 144–145
Carpeting, 56–58, 240
  acrylic, 57
  cost comparison chart, 58
  dining room, 256–257
  features, 57
  living room, 213

Carpeting (*cont.*)
  nylon, 56
  olefin, 57
  padding, 58–59
  polyester, 57
  types of, 57
Casement windows, 128–129
Cedar shakes, interior walls with, 41
Cedar shingles, 15
Cedar siding, 300
Ceiling fans, 96–97, 236
Ceilings, 25–41
  basement, 220, 222, 249
  bathroom, 197
  bedroom, 230–231, 238–239
  causes for defects in, 34
  dining room, 258–259
  exposed-beam, 35, 39
  kitchen, 176–177
  living room, 214
  plank-type, 36, 37, 38
  plaster, 28–37
    application techniques, 33–34
  sun room, 280
  tall, 31
  tongue-and-groove planking, 37–39
  vaulted, 223
  wood, 35
Ceiling tile
  applications, 33
  comparison chart, 33
  grid system, 29
  installing, 30
  samples, 31, 32
  types of, 30
Ceramic mosaic tile
  installation difficulty rating, 71
  sizes, 71
Ceramic tile, 44, 51–52, 53, 67
  in foyers, 72
  in showers, 92
  installation difficulty rating, 71
  sizes, 71
Chair rail, 121, 253–254, 259
Cinder block, on pier foundations, 4
Cinder block foundations, 5–6
  painting, 5
  parging and painting, 5–6
Circlehead windows, 130
Concrete foundations, 8–9
Countertops
  costs, 150
  kitchen, 146–151, 181–183
    costs, 150

Index

Countertops, kitchen (*cont.*)
   rolled-edge, 147
Covenants, roofing, 20–21
Cracks
   in solid concrete walls, 9
   repairing in brick foundations, 6–7
Crown molding, 122

D

Decks
   doors for (*see* Patio doors)
   existing, 306–307
   foundations, 310
   new, 307–310
Den, 167
Dining rooms, 163, 253–266
   ceilings, 258–259
   chair rail, 253–254
   doors, 260–261
   flooring, 256–258
   lighting, 99, 265
   room size, 254–256
   wall treatments, 259–260
   windows, 262–265
Dishwashers, 81
Doors, 133–139
   basement, 251–252
   bathroom, 198–200
   bedroom, 242
   brass kickplates, 161–162
   costs, 140–142
   deck or patio, 135–139
   dimensions, 136
   dining room, 260–261
   exterior, 133–135
   fiberglass, 134–135
   French, 139
   glass, 135
   gliders, 138
   hinged patio, 139
   kitchen, 179–180
   mail slots, 162
   metal insulated, 133
   passageway widths, 136
   prehung, 124
   repainting, 141
   replacing exterior, 295–296
   sliding glass, 136–138
   technical considerations, 140
   terrace, 236
   wood, 134
Dormers, false, 21–22
Double-hung windows, 129–130

Drywall, types of, 25
Ductwork
   conversion tables for
      4 to 8-inch round, 109
      9 to 12-inch round, 109
      10 to 14-inch round, 110
      16 to 20-inch round, 110
   exposed, 108

E

Electrical systems, 95–105
   amperage needed for appliances, 104
   ceiling fans, 96–97
   conductors allowed in electrical boxes, 104–105
   electrical codes, 104
   kitchen, 183
   lighting (*see* Lights and lighting)
   outlet receptacles, 96
   switch covers, 96
Electricity (*see* Electrical systems)
Epoxy paint, 48

F

Family rooms, 217–225
   basement, 219–222
      ceilings, 220, 222, 247
      flooring, 221
      lighting, 222
      walls, 221–222
   combination-type, 223–225
   flooring, 221
   lighting, 99, 222
   main-floor, 222–223
   types of, 218
Fans, ceiling, 96–97, 236
Faucets, 79
   lavatory, 90–92
   single-handle, 90, 91
   two-handle, 90, 91
Fiberglass shingles, 15
Fingerjoint trim, 120–121
Fireplaces, 168–169, 223, 235–236
Fish tanks, built-in, 237–238
Fixtures, bathroom, 197–198 (*see also* Bathrooms, bathing units; Bathrooms, bathtubs; Toilets)
Floodlights, 102
Flooring, 55–73
   basement, 221, 248
   bathroom, 187–190

Index

Flooring (*cont.*)
  bedroom, 230, 239–240
  brick, 55, 68–69
  carpeting, 56–58, 213, 240, 256–257
    acrylic, 57
    cost comparison chart, 58
    dining room, 256–257
    features, 57
    living room, 213
    nylon, 56
    olefin, 57
    padding, 58–59
    polyester, 57
    types of, 57
  ceramic tile, 66–68, 72
  dining room, 256–258
  installation difficulty rating chart, 71
  kitchen, 171–174
  quarry tile, 55, 66
  slate, 68
  sun room, 279
  tile, 71, 190, 257–258
  tile patterns, 73
  vinyl, 56, 59–65, 173–174, 188–189
    costs, 60
    cushion factor, 61–62
    features, 59
    FHA and VA standards and, 59–60
    inlaid vinyl, 63
    installation surfaces, 63–65
    no-wax, 61
    rotovinyls, 63
    solid sheets of, 62–63
    stretch-cushioned vinyl, 63
    types of, 63
  wood, 55–56, 65–66, 174, 213, 256
Flowers, 313–314
Footings, soil and, 7
Foundations, 1–9
  brick, 6–8
    bad bricks, 8
    bricks pointing up, 7–8
    cleaning, 8
    cracks in, 6–7
    soil loads, 7
  cinder block, 5–6
    painting, 5
    parging and painting, 5–6
  fixing, 1–9
  for decks, 310
  pier, 3–4

Foundations (*cont.*)
  solid concrete, 8–9
Foyers, 162

G

Garages
  home office in, 286–288
  laundry rooms in, 271–272
Garbage disposals, 80–81
Gardens, 315
  lighting in, 103
Glass walls, 280
Groundcovers, 313–314
Gutters, 300–301
  costs, 301
  life span, 301

H

Hallways, light fixtures for, 98
Hardboard siding, 13–14, 300
  horizontal wood, 14
  vertical wood, 14
  wood shingles, 14
Heat, sources of losing/gaining, 114
Heat pumps
  air-source, 115
  choosing, 116
  earth-source, 116
  surface-water-source, 115
  well-water-source, 116
Heating (*see* Heat pumps; HVAC systems)
Home office, 283–292
  access to, 284–286
  attic, 288–289
  basement, 288
  existing rooms as, 289–291
  garage, 286–288
  locations, 283–284
  requirements, 291–292
HVAC systems, 107–116
  abbreviations common to, 113
  baseboard units, 111
  ductwork conversion tables
    for 4 to 8-inch round, 109
    for 9 to 12-inch round, 109
    for 10 to 14-inch round, 110
    for 16 to 20-inch round, 110
  ductwork, exposed, 108
  exposed piping, 108–111
  heat pumps

Index

HVAC systems, heat pumps (*cont.*)
    air-source, 115
    choosing, 116
    earth-source, 116
    surface-water-source, 115
    well-water-source, 116
  new, 113–116
  outside equipment, 112–113
  radiators, 111–112
  sources of heat loss/heat gain, 114

K

Kickplates, brass, 161–162
Kitchens, 171–185
  accessories, 80, 163–164
  appliances, 183–184
    dishwashers, 81
    garbage disposals, 80–81
  cabinets, 143–146, 180–181
    choosing, 154–157
    construction features, 151–154
    costs, 146, 154–157
    features, 145
    installing, 157–158
    refacing, 144–145
  ceilings in, 176–177
  countertops, 146–151, 181–183
  doors in, 179–180
  electrical system, 183
  faucets, 79
  flooring for, 171–174 (*see also* Flooring)
  light fixtures for, 98
  plumbing, 75, 76–81, 183
  replica fixtures, 79
  sinks, 76–78
    accessories for, 80
    refinishing, 78–79
  skylights in, 177
  tile flooring, 66–67
  wall treatments, 174–176
  windows in, 177–178

L

Landscaping, 311–316
  flowers, 313–314
  garden settings, 315
  groundcovers, 313–314
  mulch, 314
  plans, 312–313
  shrubs, 313

Landscaping (*cont.*)
  sprinkler systems, 315–316
  trees, 313
  walkways, 314
Landscaping timbers, 314
Latex paint, 16, 46–47
Lattice, on pier foundations, 4
Laundry rooms, 267–273
  accessories, 166–167
  basement, 269, 271
  closet, 268–269
  clothes lines in, 271
  garage, 271–272
  in-house, 272–273
  sewing area in, 271
  types of, 267–268
Lavatories, 85–89
  drop-in, 86
  faucets for, 90–92
  integral tops, 87–89
  pedestal, 87
  rim-type, 86
  wall-hung, 86
Lights and lighting, 95, 97–104
  accent, 103
  basement, 222
  bathroom, 98, 198
  bedroom, 243
  coach-lantern lights, 101–102
  costs, 97–98
  dining room, 99, 265
  electrical codes, 104
  exterior, 100–104, 302
  family room, 99, 222
  floodlights, 102
  garden, 103
  hallway, 98
  kitchen, 98
  post lanterns, 103
  recessed, 99–100
  security, 103
  skylights, 23–24, 131–133, 177, 200, 236
  spotlights, 102
  sun room, 281
  track, 101, 102
  under cabinet, 163
  walkway lights, 103
Living rooms, 162, 205–215
  carpeting, 213
  ceilings, 214
  selecting a style, 208–212
  selecting materials for, 213–214

Living rooms (*cont.*)
  types of, 206–208
    above average formal, 208
    average formal, 206–208
    fancy formal, 208
  wall treatments, 214
  windows, 214

M

Mailboxes, 162
Mail slots, 162
Medicine cabinets, 165, 202–204
Mirrors, bathroom, 201
Molding
  crown, 122
  shoe, 118–119
Mulch, 314
Murals, 44–45, 54, 232

O

Octagonal windows, 130, 200
Oil-based paint, 16, 47

P

Padding, carpet, 58–59
Paint, 16, 241
  acrylic, 16, 47
  alkyd, 16, 47, 48
  dripless, 48
  epoxy, 48
  exterior, 46–47
  interior, 47–48
  latex, 16, 46–47
  oil-based, 16, 47
  one-coat, 48
  textured, 48
Painting, 44, 45–48
  choosing colors, 46
  cinder block foundations, 5–6
  doors, 141
Paneling, 39–40
  wainscotting, 122–124, 223, 259
Parging, cinder block foundations, 5–6
Patio doors, 135–139
  French-style, 139
  gliders, 138
  hinged, 139
  sliding glass, 136–138
Pier foundations, 3–4
  block, 4

Pier foundations (*cont.*)
  brick, 4
  facades, 4
  lattice, 4
Pine siding, 300
Planking
  tongue-and-groove, 37–39
  using on ceilings, 36, 37
Plaster ceilings, 28–37
  application techniques, 33–34
Plaster walls, 28–37, 213–214
Plumbing, 75–94
  bathrooms (*see* Bathrooms)
  faucets, 79
  fixtures, 197–198
  kitchens (*see* Kitchens)
  sinks
    kitchen, 76–80
    refinishing, 78–79
    troubleshooting problems with, 77
  sun room, 281
Polyurethane, 48, 126
Porches
  adding, 304–306
  existing, 303–304
Property deeds, covenants and restrictions, 20–21

Q

Quarry tile, 55, 66, 257
  installation difficulty rating, 71
  sizes, 71

R

R-value, 140
Radiators, 111–112
Railings, exterior, 296–297
Restrictions, roofing, 20–21
Roofing, 14–15, 17–24
  asphalt shingles, 15
  cedar shingles, 15
  color, 17–18
  false dormers on, 21–22
  fiberglass shingles, 15
  materials comparison chart, 21
  materials life span chart, 22
  old, 18–20
  property deed covenants/restrictions, 20–21
  replacing, 297–298

Index

Roofing (cont.)
  skylights in, 23–24, 131–133, 177, 200, 236
  slopes for various materials, 19
  stoop covers, 22–23
  windows in, 132–133 (see also Skylights)

S

Sealers, 126
Security, lighting for, 103
Shakes, cedar, 41
Shingles, 299–300
  asphalt, 15
  cedar, 15
  fiberglass, 15
  wood, 14
Shoe molding, 118–119
Shrubs, 313
Shutters, 301–302
Siding, 11–14
  aluminum, 11–12, 299
  cedar, 300
  comparison chart, 12, 13
  hardboard, 13–14, 300
    horizontal wood, 14
    vertical wood, 14
    wood shingles, 14
  pine, 300
  replacing, 298–300
  shingles, 299–300
  vinyl, 12, 299
Sinks
  accessories for, 80
  faucets for, 79
  kitchen, 76–80
  refinishing, 78–79
  replica fixtures for, 79
  troubleshooting problems with, 77
Skylights, 23–24, 131–133, 177, 200, 236
Slate flooring, 68
Sliding glass doors, 136–138
Soil, footings and loads on, 7
Spotlights, 102
Sprinkler systems, underground, 315–316
Staining, 44
  woodwork, 51
Stenciling, 44, 54, 232
  in kitchens, 176
Study, 167

Sun rooms, 275–281
  ceilings, 280
  flooring, 279
  glass walls, 280
  lighting, 281
  plumbing, 281
  purposes for, 278–279

T

Terrace doors, 236
Tile
  ceramic, 44, 51–52, 53, 67, 71, 72, 92
  ceramic mosaic, 71
  fancy wall design using, 170
  installation difficulty rating, 71
  patterns, 73
  quarry, 55, 66, 71, 257
  sizes of, 71
  wall, 194–195
Tile flooring, 190
  bathroom, 68
  dining room, 257–258
  kitchen, 66–67
Toilets, 84–85
  improper alignment, 88
  minimum width requirements, 87
  proper alignment, 89
Track lighting, 101, 102
Trees, 313
Trim
  clamshell, 120
  colonial, 120
  fingerjoint, 120–121
  kits for, 124
  simple boards, 120
  styles of, 120–121

U

Urethane, 48
U-value, 140
UV-blockage rating, 140

V

Vanities, bathroom, 151 (see also Cabinets)
Varnishes, 126
Vinyl flooring, 56, 59–65, 173–174, 188–189
  costs, 60
  cushion factor, 61–62

Vinyl flooring (*cont.*)
  features, 59
  FHA and VA standards and, 59–60
  inlaid vinyl, 63
  installation surfaces, 63–65
  no-wax, 61
  rotovinyls, 63
  solid sheets of, 62–63
  stretch-cushioned vinyl, 63
  types of, 63
Vinyl siding, 12, 299

W

Wainscotting, 122–124, 223, 259
Walkways, 314
  lighting for, 103
Wallpaper, 44, 48–50, 241–242
  bathroom, 193
  borders, 52
  dining room, 259
  kitchen, 174
  removing, 49–50
  surface preparation, 49
  types of, 49
Walls, 25–41
  basement, 221–222, 249
  bathroom, 40–41
  glass, 280
  partition, 291
  plaster, 28–37, 213–214
  solid concrete, 8–9
  tongue-and-groove planking, 37–39
  weathered barn board, 27–28, 223
Wall treatments
  bathroom, 191–197
  bedroom, 231, 240–242
  borders, 52
  bricks used for interior, 26–27
  cedar shakes on, 41
  ceramic tile, 44, 51–52, 53, 170
  dining room, 259–260
  kitchen, 174–176
  living room, 214
  murals, 44–45, 54, 232
  painting (*see* Paint; Painting)
  paneling, 39–40
  stenciling, 44, 54, 176, 232
  tile, 44, 51–52, 53, 170
  wainscotting, 122–124, 223, 259
  wallpaper, 44, 48–50, 52, 193, 241–242, 259

Whirlpool bathtubs, 83
Windows, 127–131
  awning, 131
  basement, 250–251
  bathroom, 198–200
  bay, 131
  bedroom, 238
  bow, 131
  casement, 128–129
  circlehead, 130
  costs, 140–142
  dining room, 262–265
  double-hung, 129–130
  kitchen, 177–178
  living room, 214
  octagonal, 130, 200
  R-value, 140
  replacing, 295
  roof, 132–133
  skylights, 23–24, 131–133, 200, 236
  technical considerations, 140
  tempered glass, 140
  U-value, 140
  UV-blockage rating, 140
  walk-out bay, 264–265
Wood
  staining, 44
  weathered barn board, 27–28, 223
Wood flooring, 55–56, 65–66, 174, 213, 256
Wood siding (*see* Hardboard siding)
Woodwork, 117–126
  chair rail, 121, 253–254, 259
  crown molding, 122
  doors (*see* Doors)
  finishes, 126
  options, 124–125
  shoe molding, 118–119
  staining, 51
  trim
    clamshell, 120
    colonial, 120
    fingerjoint, 120–121
    simple boards, 120
    styles of, 120–121
  trim kits, 124
  wainscotting, 122–124, 223, 259

Z

Zoning, covenants and restrictions, 20–21

About the Author

Chase Powers is a familiar name in the building and remodeling industry. He is a licensed general contractor with extensive experience in both building and remodeling on a residential and light-commercial level.

His work has included all types of remodeling, including attic and basement conversions, room additions, cosmetic changeovers, and many more nonstructural as well as structural changes. Kitchen and bathroom remodeling have long been his specialties. Among his other accomplishments, he has conducted home remodeling seminars across the East Coast; he brings a wealth of experience to the writing of this book.